Understanding Life in the Universe

The two most fascinating questions about extraterrestrial life are where it is found and what it is like. In particular, from our Earth-based vantage point, we are keen to know where the closest life to us is, and how similar it might be to life on our home planet. This book deals with both of these key issues. It considers possible homes for life, with a focus on Earth-like exoplanets. And it examines the possibility that life elsewhere might be similar to life here, due to the existence of parallel environments, which may result in Darwinian selection producing parallel trees of life between one planet and another. *Understanding Life in the Universe* provides an engaging and myth-busting overview for any reader interested in the existence and nature of extraterrestrial life, and the realistic possibility of discovering credible evidence for it in the near future.

Wallace Arthur is an evolutionary biologist with an interest in how life evolves on other planets, and in particular how similar the life forms that evolution produces on one inhabited planet may be to those on another. He is the author of *The Biological Universe* (Cambridge University Press, 2020) and *Understanding Evo-Devo* (Cambridge University Press, 2021).

The *Understanding Life* series is for anyone wanting an engaging and concise way into a key biological topic. Offering a multidisciplinary perspective, these accessible guides address common misconceptions and misunderstandings in a thoughtful way to help stimulate debate and encourage a more in-depth understanding. Written by leading thinkers in each field, these books are for anyone wanting an expert overview that will enable clearer thinking on each topic.

Series Editor: Kostas Kampourakis http://kampourakis.com

Published titles:

Forthcoming:

Understanding Life in the Universe

WALLACE ARTHUR
Emeritus Professor of Zoology, University of Galway, Ireland

CAMBRIDGE
UNIVERSITY PRESS

University Printing House, Cambridge CB2 8BS, United Kingdom

One Liberty Plaza, 20th Floor, New York, NY 10006, USA

477 Williamstown Road, Port Melbourne, VIC 3207, Australia

314–321, 3rd Floor, Plot 3, Splendor Forum, Jasola District Centre,
New Delhi – 110025, India

103 Penang Road, #05–06/07, Visioncrest Commercial, Singapore 238467

Cambridge University Press is part of the University of Cambridge.

It furthers the University's mission by disseminating knowledge in the pursuit of
education, learning, and research at the highest international levels of excellence.

www.cambridge.org
Information on this title: www.cambridge.org/9781009207362
DOI: 10.1017/9781009207355

First published 2023

Printed in the United Kingdom by TJ Books Limited, Padstow Cornwall

A catalogue record for this publication is available from the British Library.

ISBN 978-1-009-20736-2 Hardback
ISBN 978-1-009-20732-4 Paperback

'A fascinating overview of life on Earth and the prospects of finding parallel forms of it on habitable exoplanets. Wallace Arthur provides an engaging, yet scientifically accurate, overview of the current knowledge and what to expect from the next Copernican revolution looming on the horizon of astronomy.'

Avi Loeb, Professor of Science, Harvard University, USA

'From the origin of the universe through to the search for oxygen biosignatures on exoplanets, this book is a marvellous and broad introduction to our efforts to find out if this fascinating replicating material we call life is to be found elsewhere in the universe, and where we might find it. It will appeal as much to a professional seeking a good review as to the layperson wanting an introduction to the subject.'

Charles Cockell, Professor of Astrobiology, University of Edinburgh, UK

'Beginning with a guided tour of life on Earth, Wallace Arthur reaches out to explore the possibility of alien life deep in the cosmos. In this provocative but scientifically argued treatise, he describes what form such life might take and the technological means by which we might discover it. A thoughtful and riveting read that excites like science fiction yet rests on science.'

Addy Pross, Emeritus Professor of Chemistry, Ben-Gurion University of the Negev, Israel

'A tremendously broad and comprehensive look at the whole panoply of issues surrounding our search for extraterrestrial life. A very useful text for anyone just starting on an exploration of the possibilities of life in the universe.'

Arik Kershenbaum, University of Cambridge, UK, author of *The Zoologist's Guide to the Galaxy*

Contents

Foreword

Are we alone in the universe? Is there anybody out there? These are questions that have preoccupied humans for a long time, with answers provoking both excitement and fear. The possibility of the existence of alien life, and the consequences of encountering it, have been the topics of many popular films – in fact, science fiction is to a large extent devoted to such matters. Whereas there are no definitive answers yet as to whether there is any kind of life in the universe beyond Earth, we now know which questions we should ask and how to try to find answers to them. This has been made possible because during the last century or so our understanding of the universe, and our methods for studying it, have advanced enormously. The book you are holding in your hands is a *tour-de-force*, which summarizes current knowledge and understanding about the possibility of life in the universe. Wallace Arthur has skilfully brought together knowledge and conclusions from various disciplines to produce a concise and coherent account of what we currently know, what we might know in the future, and what we are likely to never know. The outcome is a fascinating and informative read that will take your thinking about life in the universe from speculative science fiction to the magnificent science of our times.

Kostas Kampourakis, Series Editor

Preface

The aim of this book is to give you a broad scientific base against which to consider the search for extraterrestrial life, along with the possible nature of that life and its possible distribution across the universe. As is appropriate for books of this series, I have aimed my explanations at a general readership. Hence I've tried not to assume too much. When in doubt, I've erred in the direction of including the scientific basics rather than omitting them. Thus the 'educated layperson' should be able to read the book in sequence from start to finish and make sense of it, without frequent digressions to other sources.

The book is an interweaving of three main strands, one of which is the detective work of the search itself (including space probes to local planets and the analysis of light from exoplanets), one of which is biological (with particular reference to evolution via Darwinian selection), and one of which is astronomical (including planetary science, with its key concept of habitability). The biological strand is focused on the possible *nature* of extraterrestrial life, in other words its chemical basis, mode of construction, overall body form, and so on. The astronomical strand is focused instead on its possible *distribution* across the observable universe – particularly our local part of it, which is the easiest to search.

Naturally, these two things – the nature and distribution of life – are interconnected. We can't set out to search for evidence of life in particular places without making some assumptions about what it's like. In particular, we typically search for two things – biosignatures, which indicate possible metabolizing life, and technosignatures, which indicate possible intelligent life. Arguably, the best biosignature to look for is atmospheric oxygen: this assumes that there is alien photosynthesis. And arguably, the best technosignature to look for is one that has been deliberately broadcast into space in the form of radio signals: this assumes intelligent life with an advanced technology.

Looking at the interconnection the other way round, it's unwise to speculate on the possible nature of life without some idea of where it might be found. In the twenty-first century, we're pretty sure there are no large life forms on any planet (or moon) other than Earth in the solar system; but we haven't yet ruled out the possibility of alien microbes, for example in the subsurface oceans of Saturn's moon Enceladus. However, in other planetary systems, large complex life forms may be as common as they are here on Earth, or indeed more so.

I've structured the book to reflect the interconnections between these three issues – the search for life, its nature, and its spatial distribution. So we weave our way back and forward freely between them. Also, there's much herein about life on Earth, interspersed with considerations about its possible extra-terrestrial counterparts. Note that the book isn't called *Understanding Extraterrestrial Life*. The title *Understanding Life in the Universe* was chosen instead for two reasons. First, the idea that at this stage in the game we can actually 'understand' alien life is a bit presumptuous, to say the least. Second, despite the protestations of some sceptics, life on Earth gives us clues about what we're likely to find elsewhere. The difficult thing, of course, is to distinguish those aspects of Earth life that are likely to be specific to our home planet from those that may be widely applicable across the universe.

At the end of the book are references and suggestions for further reading. These are organized by chapter. Within each chapter, they are divided according to the topics they address, given in the sequence they follow in the chapter concerned. Within each topic, sources are ordered alphabetically. I should stress that these chapter-by-chapter lists are only small samples of a vast literature that covers several branches of science, from cosmology to evolutionary biology, and several levels of detail, from books in the popular science genre to primary research papers. My criteria for inclusion of sources were varied. They ranged from citing papers that describe particularly important discoveries, to pointing you in the direction of highly readable treatments of particular topics, to including some sources that are of historical importance in our ever-evolving search for life in the universe.

Acknowledgements

I would like to thank the following for their help at various stages from inception to publication. Katrina Halliday, for the invitation to write this book for the *Understanding Life* series. Jessica Papworth and Olivia Boult, for editorial advice. Fred Stevenson, for reading the draft manuscript with an astute astronomical eye. Kostas Kampourakis, for suggesting what became the final title, for reading the manuscript, and for writing the foreword. David Catling, Mike Guiry, and Steve Selesnick, for their helpful comments on an earlier version of the manuscript. Bridget Bravo, for transforming my rough scribbles into professional illustrations. Jenny van der Meijden and Aloysias Thomas, for overseeing a smooth production process. Hugh Brazier, for clarifying unclear parts of my text at the copy-editing stage, as he has done for several of my earlier books. Many thanks to you all. The draft manuscript for a book is the individual work of the author, but the final printed product is very much the work of a team.

1 The Search for Extraterrestrial Life

Are We a Privileged Generation?

For many millennia, humans have gazed up in wonder at the night-time sky. The full panoply of the Milky Way is an awesome sight. The scale of space is immense. Is there life out there somewhere? If so, where, and what form does it take? In the space of a couple of sentences, we've already gone from generalized wonder to specific questions. The next step is from questions to hypotheses, or, in other words, proposed answers. Here are two such hypotheses that I'll flesh out as the book progresses: first, life exists on trillions of planets in the universe; second, it usually follows evolutionary pathways that are broadly similar to – though different in detail from – those taken on Earth.

Going from cosmic wonder to questions about alien life and on to hypothetical answers has been possible for a long time. But the final steps – from hypothetical answers to testing the hypotheses to arriving at an understanding of reality – have so far proved beyond humanity's grasp. That may be about to change. We may be on the threshold of being able to answer the age-old question 'are we alone in the universe?' with a resounding 'no'. But what is the basis of this claim? The short answer is advancing technology, and in particular advancing design of telescopes. We'll get to that topic soon. But first, let me reassure you that I'm not alone in my optimism about the imminent discovery of evidence for alien life.

A 2021 paper in the leading scientific journal *Nature*, by the American physicist James Green and his colleagues, began with the following sentence: 'Our generation could realistically be the one to discover evidence of life beyond Earth.' Its authors clearly share my optimism in this respect. Of course, there's a question-mark hanging over the meaning of 'our generation', as the six authors are varied in age, and their readers even more so. But that's a small uncertainty in the grand scheme of things. The key point is that many scientists suspect that the first evidence of life beyond Earth will be forthcoming on a timescale of years to decades, not centuries. Even if it takes five decades to acquire the first conclusive evidence, a reasonable proportion of those alive today (regrettably not including me!) will still be here to see it.

In the previous paragraph, I used the phrase 'many scientists'. But scientists of what type? Science is a broad endeavour, and some parts of it are more relevant to the search for life than others. However, that said, many scientific disciplines have a role to play. They include cosmology, astronomy, astrophysics, planetary science, atmospheric science, geology, ecology, evolutionary biology, genetics, molecular biology, and organic chemistry – and that is far from being an exhaustive list. The principal discipline is astrobiology, which is a mix of some or all of the above, with the exact mix depending on the astrobiologist concerned. Astrobiologists of my generation weren't trained as such, as there were no courses in the subject, so we've migrated towards it from various starting points – in my case evolutionary biology. But there's a younger cohort of astrobiologists who have indeed been trained in this discipline.

Importantly, the 'many scientists' who suspect that the first evidence of alien life will be found on the years-to-decades timescale include specialists in all of the disciplines mentioned above, and indeed others too. So this optimism isn't a passing fad in one particular narrow branch of science. Rather, it's a reasoned assessment of the stage we have reached in the process of searching for life, and where we're likely to get to with the new generation of space telescopes that are currently at various stages of planning, construction, or early operation, including the James Webb Space Telescope (alias JWST or simply 'Webb'), more on which later.

The paper in *Nature* that I mentioned above is focused on formulating a policy for reporting possible evidence of alien life. The authors highlight the complementary problems of false positives and false negatives. The former involve concluding that some observation is evidence for life when in fact it's not; the latter involve failing to detect life on a planet when it's actually there. They particularly stress their concern about the likely sensationalization in the popular media of results that turn out to be false positives, and I think they are right to do so. Their approach to avoiding – or at least reducing – such problems is to develop a scale of the quality of possible evidence of alien life, and to make clear where on this scale any particular finding falls. This way, a hint of possible life shouldn't be mistaken for a certainty. In Chapter 5, we'll apply their approach to claims of extraterrestrial life in the solar system.

In a way, I see this book as being complementary to the paper by James Green and colleagues. Those authors were concerned with understanding the nature of possible evidence for alien life in a rational framework, and preventing distortion of that evidence. I'm concerned with understanding the multidisciplinary basis of the search for life, and preventing flights of fancy that are too narrowly based. For example, a biologist might get excited about the possibility of evolution taking place on a particular planet, when astronomical considerations suggest that's unlikely – for example because of a short lifespan of its host star. Equally, an astronomer might get excited about evolution taking place very rapidly on some planets via large-effect mutations, when the biological argument against this happening isn't restricted to evolution on Earth. And some scientists (or, more likely, some non-scientists) might get excited about the possibility of life based on silicon rather than carbon, without being aware of the unlikelihood or impossibility (it's hard to say which) of having a silicon-based molecule that can rival DNA in terms of informational capacity.

A Brief History of the Search

To look at the history of our search for extraterrestrial life, and its philosophical foundations, ancient Greece is a good place to start. Most of the impressive philosophers there in the period from about 600 to 150 BCE (Before the Common Era, alias BC for Before Christ) included astronomical matters in their thoughts. They wondered very deeply about the things they saw in the night-time sky. Anaximander (c. 610 to 546 BCE) argued for a plurality of worlds rather than a single one. However, his approach was constrained by the limited knowledge available at the time on the nature of what we now know to be stars and planets. His 'many worlds' were abstract philosophical entities, not solid rocky bodies orbiting stars. Anaxagoras (c. 500 to 428 BCE) is thought to have been the first person to propose that stars are distant suns – or conversely that our Sun is a close-up star. And Aristarchus of Samos (c. 310 to 230 BCE) may have been the first person to propose a heliocentric solar system, with Earth and the other planets known at the time orbiting the Sun.

Prior to the realizations that (1) stars are suns and (2) planets orbit suns, hypotheses about extraterrestrial life couldn't be formulated in the way that's possible today. However, as early as Aristarchus's time, such hypotheses would have been possible. We don't know of any hypotheses of this kind from way back then. One of the reasons for the lack of them was the unfortunate failure of Aristarchus to persuade most of his contemporaries of the truth of his heliocentric view. It wasn't until the sixteenth century that Copernicus (1473–1543) would succeed where Aristarchus had failed, and a heliocentric view became commonplace among astronomers.

Although future historical research may turn up ideas about life being found on planets other than Earth between the times of Aristarchus and Copernicus, the earliest known hypotheses of this kind at the time of writing were those of the post-Copernican Italian priest–astronomer Giordano Bruno (1548–1600). Bruno's view was that stars were distant suns (following Anaxagoras), that each of these had planets orbiting it in the same way as does our local Sun (a generalization of the views of Aristarchus and Copernicus), and that many of these planets were homes to life. Bruno was a man before his time; the first planets orbiting other stars than the Sun – collectively termed exoplanets – weren't discovered until the late twentieth century. Unfortunately for humanity, and even more so for him personally, his free-thinking approach to both astronomy and theology fell foul of the Catholic Church's Roman Inquisition. Bruno was burned at the stake – an unimaginably cruel form of homicide – in 1600, in the Campo de' Fiori in central Rome, where a statue of him can now be found.

The first recognized telescope was made in the Netherlands in 1608, though there may have been previous prototypes. Galileo famously used a telescope to observe the moons of Jupiter orbiting the giant planet in 1610, setting him off on a course that would also fall foul of the Church, though in his case terminating in house arrest rather than execution. As far as we know, Galileo himself didn't develop hypotheses about alien life, but as telescopes evolved, other users did. The Italian astronomer Giovanni Schiaparelli (1835–1910) observed linear features on Mars that he called *canali* – which translates into English either as 'canals' (which implies human construction) or 'channels' (which doesn't). Schiaparelli wrote about a possible technological civilization on Mars, as did the American astronomer Percival Lowell (1855–1916), who was much struck by the supposed canals. Lowell published a 1906 book entitled *Mars and Its Canals,* with a 1908 follow-up, *Mars as the Abode of Life*. In the end, this whole line of enquiry came to nothing, when it was shown that the 'canals' were optical illusions.

The modern era of the search for life can be dated to around 1960, with the start of the endeavour called SETI (Search for Extraterrestrial Intelligence). A pioneering paper on the possibility of communication with intelligent life on other planets was published in 1959 by the astronomers Giuseppe Cocconi and Philip Morrison. Frank Drake and others began practical SETI work with Project Ozma in 1960. This was a search for radio signals from alien civilizations, conducted using the National Radio Astronomy Observatory, now the Green Bank Observatory, in West Virginia. Of course, the search for alien intelligence is only one part of the search for life. Nevertheless, from the 1960s onwards, both astrobiology in general and SETI in particular have evolved side by side in a much more continuous way than did the search for life before the second half of the twentieth century.

The Importance of Telescopes

It's a long way from Galileo's tiny telescope of 1610 to today's large ground-based instruments such as Chile's VLT (Very Large Telescope), and their space-based counterparts such as Hubble and Webb. I'll now look at a few selected milestones along this evolutionary journey. If you'd like a bit more detail than I can give in just a few pages on the functioning and/or history of telescopes, I can recommend Geoff Cottrell's 2016 book *Telescopes: A Very Short Introduction*.

Before examining milestones in the evolution of telescopes, it's worth reminding ourselves of why, from an astrobiological perspective, they are collectively important. Telescopes of Galileo's day were great for astronomy – especially when compared with the naked eye – but they weren't much use for astrobiology. In contrast, today's telescopes have the capability of addressing hypotheses about alien life. For example, we can now use telescopes to examine the atmospheres of distant exoplanets. We can search for signs of oxygen, which is regarded as a biosignature. In particular, it may be a signature of photosynthesis by microbes and plants. It's true that oxygen may come to exist in planetary atmospheres by abiotic means. However, as we'll see in Chapter 6 (penultimate section), it may be possible to distinguish the two types of origin of oxygen. This is a hugely exciting prospect.

So, what changes in telescope design have contributed to our current ability to look for biosignatures? Here are the main ones. First, increasing size. Telescopes are first and foremost light-gathering devices. The amount of light they can collect depends on their size – and in particular the diameter of their primary lens or mirror. The earliest telescopes were very small. Galileo's had a diameter of 3.7 centimetres, Newton's (in 1668) 2.5 centimetres. They're not directly comparable because Galileo's was a refractor (a telescope that uses lenses) while Newton's was a reflector (a telescope that uses mirrors), but they were both tiny by current standards. Today's large telescopes have mirrors with diameters measured in metres rather than centimetres. For example, the Gran Telescopio Canarias, situated on the Spanish island of La Palma, has an aperture of 10.4 metres.

The increase in telescope size has been continuous from the 1600s to the 2000s, though it has speeded up significantly since 1900. The largest telescope in the world at the end of the nineteenth century was the Leviathan of Parsonstown in County Offaly, Ireland, which had a diameter of 1.83 metres. This size was exceeded in 1917 with the construction of the Hooker Telescope on Mount Wilson in southern California, with a diameter of 2.5 metres. So the last 100 years have seen the largest increase – more than 8 metres between the early twentieth and early twenty-first centuries.

However, these measurements in metres don't quite suffice to explain just how much more powerful modern telescopes are than their ancient forerunners. The light-collecting capability of a telescope scales not with the diameter of its primary mirror but with its surface area. So, in terms of functionality, a 10-metre mirror is not 10 times as good as a 1-metre mirror; rather it's better by a factor of 100. Given that Galileo's tiny telescope allowed him to see the moons of Jupiter, and discern individual stars in the 'milk' of the Milky Way, the capabilities of today's telescopes are truly amazing.

But, for astrobiology, size is not enough. There are several other developments that we need to understand. First, so obvious that it hardly needs to be said, the advent of photography in the 1800s, and its later coming together with telescopes in the form of astrophotography. Second, Earth's atmosphere is a major cause of distortion in the business of seeing clearly into space. As is well known, stars don't twinkle – they only seem to do so because of atmospheric effects. The main exoplanet-detecting telescopes have been placed above the atmosphere so that this distortion is avoided. These are the Kepler and TESS space telescopes (more on which in Chapter 6). The first is named after the German astronomer Johannes Kepler (1571–1630), who formulated the laws of planetary motion; the second is an acronym of Transiting Exoplanets Survey Satellite, with 'transiting' referring to the movement of a planet across the face of its host star, as seen from our perspective.

As well as making a telescope bigger, enabling it to take images, and in some cases launching it into space, another development we need to consider is the range of wavelengths it can see. 'Light' is a slippery term. It's sometimes used for the part of the electromagnetic spectrum we can see, though this is better described as 'visible light'. Alternatively, it can be used more broadly to refer to the whole of the spectrum – as in the distinction between the speeds of light waves and sound waves. Early telescopes were designed – understandably – to see visible light. Today's telescopes are much more varied in this respect. Some can see in the ultraviolet range, some in the infrared range, some can see radio waves, and some can see across boundaries between our quasi-arbitrary sections of the spectrum.

One example of 'seeing across boundaries' is the James Webb Space Telescope, launched in December 2021. Another will be a future space telescope, which, at the concept stage, was called LUVOIR, an acronym of Large Ultra-Violet, Optical, and Infra-Red (telescope). This design survived NASA's 2020 decadal survey, which produced its report in late 2021 due to the Covid pandemic. However, it was scaled down in size for reasons of cost.

The primary mirror will be of a smaller diameter than was originally envisaged. When constructed, the diameter will be about 6 metres – still a considerable size for a space telescope. Its final name will probably be different. In the immediate aftermath of the decadal survey, some astronomers jokingly suggested simply losing the L (large) and calling it UVOIR, but a better name will almost certainly be found. The ability to see across the boundary between visible and infrared light that characterizes both James Webb and LUVOIR means that they will be ideally placed to see oxygen biosignatures.

This point brings us to the other important development in telescopes that we need to examine: the use of a technique called spectroscopy. Here's the basic idea behind this technique, as applied in space telescopes for astrobiological purposes. You look at light from a distant source – say a star that has an orbiting planet. In fact, you look at it twice, once when the planet is behind the star or a long way off to one side, and once when it is transiting – i.e. in front of its star from our perspective. You compare graphs of amount of light received versus wavelength for the two occasions (Figure 1.1). Dips in the amount of light at particular wavelengths when the planet is in front of the star are interpreted as resulting from absorption of light by gases in the planet's atmosphere. Since different gases have different characteristic patterns of absorption, you can identify them.

There's a final twist to this story. The effects of a planetary atmosphere are tiny compared with the immense blast of light that hits our telescope from the fiery furnace of the planet's host star. A better approach would seem to be this: look not at the star with the planet in front of it but rather at the planet itself. There's a big problem here that at first seems insurmountable but is not. Even when the planet is a long way off to the side of the star, starlight dominates what you see, and in any event planets don't produce light of their own. The solution to this problem is twofold. First, where possible, blot out the light from the star using some form of starshade. Second, observe in the infrared range because, while planets don't emit their own visible light, they *do* emit in the infrared; in contrast, starlight peaks in the visible region and declines in the infrared. Looking directly at planets in this way is called direct imaging.

Seeing (or 'listening') in the radio range is also important – especially from a SETI perspective. As well as searching for biosignatures that could come from very

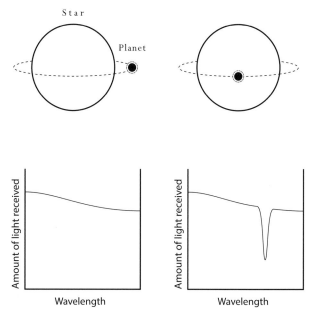

Figure 1.1 Detecting gases in exoplanet atmospheres. Left: amount of light received by a space telescope when the planet is 'off to one side' from our perspective. Right: amount of light received when the planet is in a direct line between its host star and the Earth. Note the dip in the amount of light received at a particular wavelength in the right-hand panel. If instead of looking at this graph you were looking at the light itself, after it had been split into its various colours by a prism, you would see a dark band at the wavelength (= colour) concerned. Such a dip (or band) is interpreted as the result of absorption by a particular gas in the planet's atmosphere, and is known as an absorption band. Since different gases have different patterns of absorption centred on different wavelengths – in other words different 'signatures' – they can be identified. Some gases, notably oxygen, are regarded as *bio*signatures. As we will see later (in Chapter 6), the idea of a single absorption band is a simplification; each gas typically has several of them.

simple forms of life such as bacteria, we also look for technosignatures – those that could only come from an advanced civilization. If such a civilization decides to broadcast its existence in the way that humans have done since the middle of the twentieth century, it will probably use radio signals to do so. Thus, radio telescopes come into their own in the search for intelligence in the universe. Well-known radio telescopes include Jodrell Bank in Cheshire, England, and the Arecibo telescope in Puerto Rico, the latter now lamentably defunct following damage sustained in 2020. It gave its name to the famous Arecibo message broadcast by it into space in 1974. There are also large *arrays* of radio telescopes – for example ASKAP (Australian Square Kilometre Array Pathfinder), located about 700 kilometres north of the city of Perth, which started observing the sky in 2012.

A Working Definition of Life

We are about to *feel* the multidisciplinary nature of the search for extraterrestrial life that I mentioned at the outset, because we are about to move from the astronomical realm to its biological counterpart. In particular, we are going to move from the telescopes used in the search for habitable planets – and ultimately the life that's on some of them – to the search for a workable definition of life. How can we search for life if we can't define it? And yet the task of defining life is not an easy one – even in the restricted context of planet Earth (are viruses alive?). It's more difficult again to attempt a definition that will work in a cosmic context. But an imperfect definition is better than nothing – so here goes.

One way of looking at a definition is as an abstract generalization about a class of entities that's based on prior knowledge of many *concrete individual entities* of the sort concerned. So, let's start with a list of 'concrete' living entities – organisms – on our home planet. Not a complete list, of course, for that would include more than a million names. But we don't need a complete list, nor even a particularly lengthy one. A short one will suffice, and indeed is preferable, in terms of not drowning in detail. Here it is: humans, snails, whales, trees, toadstools, and bacteria.

A definition of 'X' – in this case 'life' – only works if we also have entities that we know are 'not X', which in this case means 'not life', or alternatively 'inert objects'. Here's a short list of such objects, all of which are generally agreed *not* to be living things: rocks, clouds, cars, smoke, and water. Of course, some of these entities might *contain* life forms – for example a cloud might contain airborne microbes – but that's not the same thing as a cloud actually being a living organism itself. And in one case – cars – they're *made by* life forms, but again that's something different.

So, the question is this: how do we word an abstract definition of life so that it includes all of the things that we think of as living and none of those that we think are not? There are different ways of doing this, and I have no intention of surveying them all. That would be both boring and pointless. Instead, I will focus on the definition that I think works best. I call it the RIM definition of life. These letters stand for the following three attributes of life forms: Reproduction, Inheritance, Metabolism. So, an entity that has all three RIM attributes is alive, while an entity with none of them is not. We'll get to the annoying grey area in the middle shortly.

Now is the moment for a warning about the risk of infinite regress. To define life, it's necessary to define reproduction, inheritance, and metabolism. This could lead to a need to define words that occur in these definitions, and so on indefinitely. You'll be pleased to know that I'm absolutely *not* going to go down that road. However, I *am* going to briefly consider each of the three components of RIM. Since these three things are crucial in allowing us to distinguish life forms from inert entities – potentially throughout the universe – it simply won't work if we're fuzzy about their meanings.

This point was brought home to me when I was reading the *Very Short Introduction* to astrobiology written by the planetary scientist David C. Catling in 2013. With regard to metabolism, Catling says the following: 'All life metabolizes but so does my car.' I don't agree with him, but I see what he means. The burning of a fuel such as petrol powers a car in much the same way as the burning of a fuel such as glucose powers a living organism. So Catling is right to see a parallel. But he's wrong – in my view – to equate automotive and biological 'metabolism'.

I have an instinctive feeling that the two are different in some fundamental way, and perhaps you do too. But what is it? Of course, the fuels are different, but that's not the best approach to this issue, because we can't guarantee that the fuel of alien life is glucose. Here's a better approach. A life form metabolizes right from the start; a car doesn't. An organism is not simply a collection of cells, membranes, vesicles, filaments, and so on, that is inert until some point in its life when it gets fired up with an infusion of glucose or some other fuel. In contrast, a car is an object of precisely this latter sort. It's a collection of metal and plastic parts that gets built in stages and hasn't a whiff of 'metabolism' until after it's complete – or has reached its 'adult' stage, to stretch the use of that term – and someone turns on the ignition.

So, to be a life form from a metabolic perspective you need to be metabolizing right from the start. Even the very first cell of a new human or other animal – the zygote – metabolizes. An organism is an *intrinsically* metabolizing thing. Its default state is to metabolize. However, to call life forms *perpetually* metabolizing things would be a step too far. We're all familiar with hibernation – in hedgehogs, for example. This is a state of reduced metabolic activity in cold weather. And there's the related phenomenon of aestivation – reduced metabolic activity in hot dry weather, as is found in some Mediterranean snails. But these two states are eclipsed by a more extreme one called cryptobiosis ('hidden life'), in which metabolism may be reduced to zero, or very nearly so.

The best example of animals exhibiting cryptobiosis are the tiny invertebrates called tardigrades, or water bears. These have played an interesting role in astrobiology, as we'll see later. So it's appropriate that they should feature in our discussion here of the definition of life. When environmental conditions become too extreme for a tardigrade to maintain normal, active, rapidly metabolizing life, it converts itself into a static, low-or-zero metabolizing structure called a tun. Both this state and the normal active state are shown in Figure 1.2. When conditions improve again, it can convert back from a tun into an active, mobile, metabolizing tardigrade. But this fact doesn't really cause us difficulties of definition. As long as we stick with a view of organisms as 'intrinsically metabolizing' rather than 'perpetually metabolizing', then we're ok.

That was a quick look at 'metabolism'. Now, what about reproduction? Cars don't reproduce. Neither do rocks or clouds of smoke. Organisms do. Perhaps this is more straightforward, then, than the case of metabolism? Well, not entirely. Regrettably, not all adult humans can reproduce. But those who can't are no less alive than those who can. In bees and ants, there are whole castes that can't reproduce – workers, for example. These are no less alive than reproductive drones and queens. Also, cases are known where two species occasionally interbreed, producing offspring that are typically sterile – such as

Figure 1.2   Two forms of the tiny invertebrate called a tardigrade or water bear. Upper right: how the animal looks under benign environmental conditions, when it is active. Lower left: a tardigrade tun, which is a resistant, non-metabolizing form that the animal converts itself into under extreme environmental conditions such as very low temperatures or lack of water. If conditions become less extreme, the tun can revert to the active form.

the mule, a hybrid between a horse and a donkey. So, we have to use reproduction as a characteristic of a *species* rather than of every individual, in our attempts to draw a line between life and non-life.

Now we come to the third attribute included in the RIM definition of life – inheritance. This applies to all life forms on Earth, though in different ways to some than to others. We humans are most familiar with the sort of inheritance that is associated with sexual reproduction, in which, as we can plainly see, offspring tend to resemble one or both parents more than they do other members of the overall population. Most other animal species have sexual reproduction, so the same phenomenon applies, even though, for example, we humans find it hard to distinguish degrees of similarity among different individual houseflies.

Inheritance still applies in the world of asexual reproduction, which is much more common in bacteria and other microbes than it is in animals. A population of a particular kind of bacterium that reproduces in this way can be thought of us a collection of clones. Each parent, on its own, produces progeny that are genetically identical to itself, so the within-clone variation is zero at the level of the gene (unless a mutation occurs), and may be quite small at the level of the organism. In contrast, the between-clone variation is greater. Since clones are to all intents and purposes families, the situation regarding inheritance is broadly the same. Finally, we must be equally careful in deciding what entity to apply inheritance to, as we were in using the term reproduction. Inheritance applies to the relationship between generations, not to individual organisms taken alone. Of the three RIM attributes, metabolism is the only one that can be applied to individuals considered in isolation from other members of their family, population, or species.

In summary, the RIM definition of life can be constructed something like this. An entity is alive if (1) it has intrinsic metabolism; (2) it belongs to, or is descended from, a group (species) that reproduces sexually, asexually, or through a mixture of the two; and (3) there is inheritance of organismic features from one generation to the next. If it has none of these features, it is not alive. Now we get to the grey area mentioned above. What happens if an entity has one or two of these attributes, rather than zero or three?

I'm quite confident that most entities on planet Earth can be classified as living or non-living on the basis of the RIM definition, as given above. However, there is an important exception: viruses. Are these entities tiny subcellular organisms, or very large and complex, but non-living, molecules? Most people think of them as being alive, but is this intuitive feeling justified? Viruses have no metabolism. At its simplest, a virus is a piece of hereditary material made of nucleic acid (sometimes DNA, sometimes RNA) wrapped in a protein coat. Its size is orders of magnitude less than that of a single cell. Viruses operate by entering host cells and hijacking their molecular machinery to make offspring viruses.

It's clear from this brief description that viruses have the R and the I of RIM, but not the M. So they possess two out of the three attributes of life. However, this statement has to be qualified in an important way. Viruses can't reproduce outside of a living cell of their host organism, be it an animal, a plant, a fungus, or a microbe. So their reproduction, and hence their inheritance, is not intrinsic. Perhaps we could conclude that they score half, half, and zero on the RIM scale, thus giving them an overall score of one rather than two. Whatever exact score we give them, the conclusion is inescapable: viruses inhabit a grey area between life and non-life.

A Working Definition of Intelligence

The searches for alien life in general, and intelligent alien life in particular, are very different endeavours. We've already seen that they involve different wavelengths on the electromagnetic spectrum, and consequently different designs of telescope. Searching for signatures of oxygen is centred on the visible and infrared parts of the spectrum. In contrast, searching for the technosignatures of intelligent civilizations is centred on longer wavelengths – those of the radio range. I'm going to use radio as a pragmatic way of defining intelligence from a SETI perspective. But let's first look at the span of intelligences among animals on Earth from a different perspective.

Intelligence is a continuous variable, not a binary one. Life forms here, and probably elsewhere too, show a wide range of behaviours that can be interpreted as indicating a wide range of intelligence underlying them. The construction of a scale for the measurement of intelligence is notoriously difficult, even within the confines of a particular species – humans. How intelligent someone is cannot be completely captured on a single scale; intelligence is a multidimensional property. Paradoxically, though, in a multi-species context the job actually gets easier rather than harder. The reason for this is that the differences we're dealing with are much greater.

A scale for intelligence in the context of life on Earth starts from zero and extends indefinitely upwards. Organisms that lack nervous systems are deemed to have no intelligence. These include all microbes and plants, as well as the most primitive of animals – sponges and a little-known group called placozoans (which translates as 'flat animals'). Slightly more advanced animals such as jellyfish have diffuse nervous systems but lack brains, and don't really exhibit any behaviour that merits the description intelligent. There are many animals that have small brains, including insects and arachnids. These typically have rather fixed, instinctive behaviour patterns, but some seem more intelligent than others. Jumping spiders – for example those of the genus *Portia* – have been put forward as the most intelligent arthropods, owing to their complex patterns of hunting behaviour.

The most intelligent invertebrates of all are octopuses; there's a great book about octopus intelligence by Peter Godfrey-Smith, called *Other Minds*. Many studies on these wonderful creatures, both in the lab and in the field, have illustrated complex behaviours that are clearly learned rather than instinctive. Octopuses exceed many vertebrates in intelligence. However, some groups of birds seem to be particularly intelligent – notably crows and parrots. And many mammalian groups show high intelligence, especially cetaceans (dolphins and whales) and primates. Among the primates, the great apes stand out, and among these, of course, humans represent an apex.

Rather than pretend that it's possible to create a cross-species numerical scale of intelligence, let's simply acknowledge five broad-brush levels: zero (e.g. sponges), little if any (e.g. worms), some (e.g. jumping spiders), a lot (e.g. octopuses, crows, most mammals), and 'high' (humans). Counterparts of these may well eventually be found among the life forms of other planets, though there may also be a further category ('very high') on some of them, the nature of which we can only dimly conjecture.

Now let's poke further into the 'high' category here on Earth. Human tool-making is thought to have originated about three million years ago – in the form of stones fashioned for hammering and chopping. That was a significant milestone. Wheels and writing – two further important milestones – were invented thousands rather than millions of years ago. Telescopes were invented a few hundred years ago – around 1600, as we've already seen.

And radio transmission was achieved in the years around 1900, with the Italian scientist Guglielmo Marconi being a key figure.

Here's an important caveat. The pattern of increasing sophistication of human inventions over time can't be mapped directly to a pattern of increasing intelligence in time. Human brain size reached its present level more than 100,000 years ago. Brain size is a very blunt instrument, but even when more subtle aspects of the brain, such as its shape, are taken into account, the brains of members of our species – *Homo sapiens* – seem to have reached something resembling the present-day condition by about 50,000 years ago. Not only that, but the Neanderthals alive at that time had brains that were at least as big as ours.

So humans of 50,000 years ago probably had an intelligence level not dissimilar to that of an average human alive today. Marconi's ability to send radio signals across the Atlantic in the early twentieth century was probably not so much a consequence of human evolution in the last 50,000 years as it was a consequence of being able to build on a wide range of discoveries and inventions that had taken place in previous history, and on the prediction of radio waves by the Scottish physicist James Clerk Maxwell in the 1860s.

It's important to keep in mind this point that the mapping of the level of technology of a society to the level of intelligence of its members is complex. However, it's also important to be pragmatic when trying to come up with a working definition. I'm going to use the ability to send and receive radio waves as indicating that a certain threshold of intelligence has been reached, because searches of a SETI type are largely conducted using radio, and the only beings we'll detect by conducting these searches are those who can send out powerful signals (radio or possibly laser) into space. The relevant intelligence threshold may have been reached thousands of years earlier, but that's not important from a practical perspective.

Two Key Questions

We've now considered what we're looking *with* – telescopes plus various associated techniques including spectroscopy. And we've also thought carefully about what we're looking *for* – both life in general, as defined by reproduction, inheritance and metabolism, and intelligent life with advanced technology in particular, as defined by the ability to send and receive radio signals. Our final job, before venturing out into the cosmos to take its measure, is to be clear about the key questions we're asking in our endeavour to understand what has sometimes been called 'the biological universe'. There are two of these questions, or perhaps more accurately two groups of questions, as follows.

First, a group that I think of as being about the *geography* of alien life. These are questions about distribution in space. For example: Is life only found on planets orbiting stars, or might it be found in other places too, such as rogue planets (which don't orbit anything) or moons, which orbit planets? Where is the closest life – and the closest intelligent life – to us? How many inhabited bodies are there, at various levels of spatial scale from the solar system through the galaxy to the universe as a whole? Are there planetary systems in which more than one planet is inhabited?

Second, a group of questions that I think of as being about the *biology* of alien life. These are questions about the nature of life forms, wherever they are found. For example: Is life usually (or always) based on carbon? Is it usually dependent on water? Is life usually made of cells or cell-like units? Are large life forms common, and do they usually arise from the evolution of multicellularity rather than by some other means? Are there features of large multicellular life forms that are to be expected to occur widely across multiple planets, for example skeletons? How does intelligent life arise from unintelligent beginnings, and are there parallels in the evolution of intelligence from one planet to another?

Although these two groups of questions are reasonably distinct from each other, they are not without overlap and interaction, as I mentioned earlier. For example, by now we are fairly sure that, Earth aside, there are no multicellular life forms in the solar system, even though the possibility of microbial life having evolved on some local body – for example Jupiter's moon Europa or Saturn's moon Enceladus – remains a possibility, as we'll see in Chapter 5. In contrast, it seems highly probable that there are plenty of rocky bodies scattered across our galaxy that have large life forms as well as microbes. Some authors have argued against this – notably the American scientists Peter Ward and Donald Brownlee in their 2000 book *Rare Earth*. These authors argued that Earth possessed a combination of unique features that was unlikely to be replicated elsewhere. However, that was before the discovery of Earth-like exoplanets, and their argument no longer seems tenable.

As we proceed, these two groups of questions – about the geography and biology of alien life – will always both be at least in the background. In particular chapters, one may be more to the fore than the other. For example, Chapter 2 will focus largely on the places where life might evolve, while Chapter 3 will concentrate on the nature and evolution of life on Earth, and possibly elsewhere. As the book progresses, I'll try to tie the two groups of questions – and possible answers – together. Ideally, we'll end up with a pretty good idea of the most likely abodes for alien life and the nature of that life, at least in general terms.

2 Where in the Universe to Look?

What Is the Observable Universe?

To look for life in the universe beyond Earth, we need to understand what is meant by 'universe', just as we need to understand what is meant by 'life'. In the end, we can probably ignore most of the universe and focus our search in some very specific places. However, those places are best understood against a backcloth of what can be called – it's an understatement really – the big picture.

The universe – alias the cosmos – is essentially 'everything, everywhere'. This includes matter, energy, and the space they travel through. But part of the universe is forever hidden from our view. We acknowledge this by using the term 'observable universe'. However, this term requires careful consideration, because it's easily misunderstood. The best way to look at it is to start from the perspective of history – specifically, the observable universe as it was in Galileo's time and as it is in our own. When the term is used correctly, the observable universe is exactly the same now as it was then. This fact, combined with the dramatic evolution of telescopes over the last four centuries, which we looked at in the previous chapter, makes clear that 'observable universe' is a theoretical notion; it's not a practical one that's dependent on our technology. The fraction of the universe that Galileo could see with his primitive telescope was much smaller than the fraction that can be seen by Hubble, or other advanced telescopes of today. We could call that fraction the 'observed universe', and it's increasing all the time. However, it will one day stop increasing, regardless of how advanced our telescopes become. This will happen when the 'observed universe' comes up against the limit of its 'observable' counterpart. That day will not be too far into the future.

Astronomical distances can be measured in light years. If you already know what these are, please skip this paragraph. One light year is the distance light travels in a year – an Earth year, that is (we should acknowledge that 'a year' varies from one planet to another). This distance has long been known with a high degree of accuracy. Rounded to the nearest thousand kilometres or miles, the distance that light travels *per second* is 300,000 kilometres or 186,000 miles. When we convert this to *per year*, and round it a bit more to give just a rough approximation, we find that light travels almost 10 trillion kilometres or 6 trillion miles in a year. So, when I talk about some particular number of light years below, you can see how to convert it into more everyday units of distance. However, such conversions produce very unwieldy numbers that are way beyond trillions; this, of course, is why we use light years for astronomical purposes.

We don't know exactly how far Galileo could see in 1610 with his 3.7-centimetre refractor telescope. However, we can have a good guess by estimating how far it's possible to see with one of today's small refractors. The answer is 'a few million light years'. In fact, someone with good eyesight looking at the night sky from a location remote from light pollution can see between 2 and 3 million light years – the faint fuzzy patch of Andromeda, the nearest galaxy to the Milky Way, is at that distance. Now compare those figures with another – the distance to the furthest galaxy confirmed so far, which goes by the code name of GN-z11 and is about 32 *billion* light years away from Earth. If you want to know where it is in the night sky, it's in the constellation of Ursa Major (the Great Bear), though of course you can't actually see it there. It was discovered in 2016 by a team analysing data from the Hubble and Spitzer space telescopes. The radius of the observable universe is about 46.5 billion light years, for reasons we'll get to in a moment. Now that we can see an object that's 32 billion light years away, we can see more than two-thirds of the maximum distance that anyone will ever see. And this fraction will keep rising. Indeed, at the time of finalizing this chapter, the discovery of a galaxy that's probably over 33 billion light years away has just been announced; and it's likely that this galaxy (called HD1) will soon be confirmed, in which case it will replace GN-z11 as the furthest known galaxy in the observable universe.

There are two initially strange-looking things about the radius of the observable universe: first, it's centred on Earth, which seems somewhat anti-Copernican; and second, it seems too big. Let's take these in turn.

Of course, science hasn't abandoned Copernicus. The only reason that our observable universe is centred on Earth is that it relates to seeing things from here. If those Martian canals had been real, their builders would have had an observable universe that was different from ours, though not by much, because the distance from Earth to Mars is tiny in the grand scheme of things – much, much less than a single light year. However, if there are intelligent life forms on a planet somewhere in that distant galaxy GN-z11, the centre of their observable universe is shifted billions of light years relative to ours. They can potentially see many things that we can't, and vice versa. There is only one universe (probably!), but there are many observable universes – one centred on each possible viewing point.

Now let's confront the problem that our estimate of the radius of the observable universe may seem a bit on the high side. A common mistake is to assert that this radius is 13.8 billion light years rather than 46.5. If you scour the web, you can even find images of signs in museums that feature this assertion. The reason it's wrong is that it fails to take into account the fact that the universe is expanding. If the universe had magically sprung into being in its present form 13.8 billion years ago and remained unchanged ever since, the most distant galaxies we could see would be 13.8 billion light years away. A galaxy that was 14.8 billion light years distant wouldn't be visible until another billion years had elapsed. Given Einstein's universal speed limit for light (and every other moving ray or object), it is logically impossible to see more than 13.8 billion light years in a static universe that's 13.8 billion years old.

But the universe isn't static – it's expanding. This takes the form of space itself expanding, rather than galaxies moving away from each other in a fixed space. Expansion has been happening – albeit at a variable rate – ever since the Big Bang. Here's an example of what this means for light travelling from one place to another, for example from a distant galaxy to Earth. Consider again that hypothetical galaxy at a present-day distance of 14.8 billion light years, but this time consider it against the background of an expanding rather than a static universe. When light from that galaxy began its journey, the universe was much smaller than it is now, so the distance it needed to travel to get here was much shorter. However, in a sense the goal posts moved as it travelled. How far they've moved for light coming from this or any other galaxy is hard to work out, given that the rate of expansion of the universe isn't constant (it's currently accelerating). But it's not impossible; detailed calculations have led to the figure of 46.5 billion light years. This means that a galaxy that's currently at a distance of 50 billion light years will never be seen by humans, however amazing the telescopes of the future become.

Understanding what's meant by 'the observable universe' is important, because it enables us to divide the overall universe into two parts – the one we can potentially investigate and the one we can't. Unfortunately, we don't know the relative sizes of the two. If the overall universe is infinite in space, despite being limited in time, then the observable universe is minuscule compared to its non-observable counterpart – if a comparison of a finite entity with an infinite one is even meaningful. On the other hand, if the overall universe is finite, then perhaps we can see a large chunk of it. But in the end, the difference between these two scenarios is academic. What's important to the search for life is that we have no choice but to restrict our efforts to the domain from which we can gather information – the realm of the potentially observable. In fact, for practical purposes in the twenty-first century search for life, we must be content with a much more restricted focus. To understand this, we need to know more about the way things are structured *within* the observable universe, which means probing into galaxies, stars and planets. Let's take them in that top-down order.

Cosmic Structure: Galaxies

At a high level of spatial scale, the observable universe is dominated by those cities of stars that we call galaxies. There are thought to be about a trillion of them. They're not evenly distributed – they are arranged in clusters and superclusters, separated by great voids. There are four main classes of galaxies, delineated by their overall shapes: elliptical, spiral, lenticular (lens-shaped), and irregular (Figure 2.1). Our own galaxy, the Milky Way, is a spiral one. Spirals are thought to make up about a quarter of all galaxies. Ellipticals are commonest, making up more than half of the total.

As well as variation in shape, there is variation in size. This can be considered in two ways – number of constituent stars or diameter measured in light years. The Milky Way is a middle-sized galaxy, with somewhere between 200 and

Figure 2.1 Classification of galaxies, based on a scheme devised by Edwin Hubble. There are four main classes: elliptical, spiral, lenticular, and irregular. Top: face-on views of these four. Bottom: side-on views. There is much variation in both size and shape within each class. The biggest galaxies are often referred to as giants, the smallest as dwarfs. Spirals can be subdivided into barred and unbarred, which refers to the presence or absence of a central bar; the Milky Way is a barred spiral. Both of these sub-classes have spiral arms.

400 billion stars, and a diameter of about 100,000 light years. At the low end, there are dwarf galaxies with star numbers that are less than a billion; at the high end there are giant ellipticals that have trillions. In general, size variation is greatest in the elliptical category.

We've already seen that the search for alien life is restricted to the observable universe. Now we ask: to what subset of this should we restrict our search, in terms of galaxies? The answer depends on the nature of the search in the following way. If we're interested in intelligent life, we can either search for unsolicited incoming radio signals or send out our own and hope to receive replies. Sometimes these two activities are classified as SETI and METI respectively, with the M standing for Messaging. Alternatively, sometimes they are referred to as passive and active SETI. Whatever we call them, they don't have counterparts in the search for non-intelligent life. If we look for the signature of oxygen in a planetary atmosphere, there's no signal being deliberately sent out by either party.

In our own METI activities, there's little point sending out signals to galaxies beyond the Milky Way, because it would take millions of years to get a reply. On the other hand, if we're searching for signals sent by intelligent civilizations elsewhere, there's no need to restrict our scope in this way. Well, no philosophical reason anyhow, in that receipt of a radio signal that was clearly broadcast into space by intelligent alien life forms would be equally amazing wherever it came from. But there's a practical reason. Signals tend to decay with distance of travel (or more precisely with the square of the distance), and thus the further away their source the harder they are to detect. So, realistically, we're more likely to be able to detect signals sent from within the Milky Way than those sent from further afield.

In fact, for many practical purposes, even the Milky Way is unrealistically large as a search field. In SETI, the search is usually for signals that were deliberately beamed into space. The intelligent life forms sending them might make them very powerful in the hope of maximizing their detectability. But photosynthetic life forms aren't characterized by a sense of purpose – well, at least not here on Earth, and perhaps not elsewhere either. The amount of oxygen they produce isn't planned or deliberately maximized. It's likely to be a weak atmospheric signature that's hard to detect. The more distant the planet we're looking at, the harder it will be, other things being equal. It isn't easy to specify a particular threshold distance to which we should always restrict our search. However, a few hundred light years would be a reasonable yardstick, and a few thousand a very optimistic one.

Using these figures, we come to the realization that for many practical purposes our search isn't restricted just to our home galaxy, it's restricted to our local part of it. But what's that? Recall that the Milky Way is a spiral galaxy. It has several spiral arms, some longer than others. Two of the longer ones are the Perseus and Sagittarius arms; running from one of these to the other is the Orion arm – sometimes appropriately called the Orion spur, given its comparatively short length (Figure 2.2). Our solar system is located on this spur, about halfway between neighbouring parts of the two longer arms. The spur is about 10,000 light years in length, 3000–4000 in width. One pragmatic view of the limit of our search area with present technology is a sphere of diameter about 3000 light years – a subset of the Orion spur centred on the Earth. This is

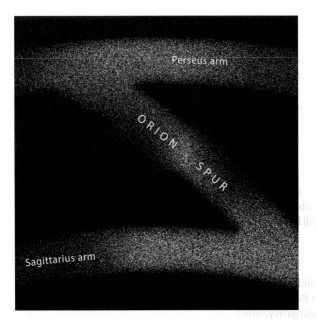

Figure 2.2 Our home region in the Milky Way galaxy. The solar system is located in the Orion spur, which runs from the inner Sagittarius arm out to the Perseus arm. In general, both arms (longer) and spurs (shorter) are imperfect structures without clear boundaries; stars gradually thin out towards their edges. Continuing down from the bottom of the zone shown leads to the galactic centre, while continuing up from the top leads to intergalactic space.

truly a tiny fraction of the observable universe, and yet it's a phenomenally large volume of space, containing many millions of stars.

Is there life on planets orbiting some of these 'local' stars? Almost certainly, given their huge numbers. Is there intelligent life on at least one of them? Probably, though that's a harder call; we'll revisit it in Chapter 8. Now for a different kind of question. How can we link our Orion-spur sphere with what we see in the night-time sky? Up to now, this exercise has been largely theoretical. That sphere we've been discussing is a very abstract thing. But we would like to know how it connects with reality in the form of the stars we can actually see. In fact, many of the stars that look bright to us are less than 100 light years away. Sirius is 8 light years from Earth; Vega, 25; Arcturus, 37; and Capella, 44. Some, however, are well beyond this limit. For example, the famous Betelgeuse ('Beetlejuice') is about 550 light years distant. And a few are beyond 1000; Deneb (the brightest star in Cygnus the swan) is about 2500 light years from Earth, give or take a fairly wide margin of error, and thus approaching the edge of our sphere.

The number of stars that can be seen with the naked eye varies a lot, depending on the location and the eyesight of the person who's looking up. For someone with average eyesight in an urban location the number will be tens, or hundreds at the very most. None of these are outside of our sphere. For someone with good eyesight in a remote location like the Scottish Highlands or the Australian outback, it's thousands rather than hundreds. In such a context, some stars will be outside our sphere, but only a small percentage of them. So there is a reasonably good connection between the conceptual sphere and the practical business of stargazing. But always beware of weasel words. The word I'm referring to is the 'reasonably' in the previous sentence. I should mention two things it hides. First, many of the stars within 3000 light years of Earth can't be seen with the naked eye, because of intrinsically low luminosity. Second, the Andromeda galaxy, which is about 2.5 *million* light years away, can be seen with the naked eye; however, its individual stars cannot.

Stars: The Powerhouses of Life

In my view, stars are essential for life, and yet they're completely uninhabitable. However, although this view is common, it's not universal. Both parts of it have been challenged. Some authors enthuse about the possibilities of life on 'rogue planets' – those dark worlds found alone in deep space, not orbiting any star. In their 2019 book *Imagined Life*, the American astronomers James Trefil and Michael Summers argue the case for life being able to originate and survive in these seemingly inhospitable places. If they are right – I suspect not – then a star isn't essential for life at all. Two other American astronomers, Luis Anchordoqui and Eugene Chudnovsky, challenge the assertion that stars themselves are uninhabitable, and argue for the possible existence of life forms in the interiors of stars, where temperatures reach millions of degrees.

The suggestion of such life at first sight seems ridiculous to most scientists. However, we should remember the importance of definitions. If life is defined more broadly than I define it here (the RIM definition of Chapter 1), then it may be found in a broader range of places than those I'll consider shortly. The proposal of life in the Sun, or in stars in general, relies on abandoning metabolism as one of the criteria for life, and focusing mainly on reproduction. Personally, I don't think that this sole criterion – or any other single measure – works to define life. Rather, I think a combination is necessary.

Setting aside the views taken by these two pairs of astronomers, stars are essential to life but not homes to it. They are the central bodies about which the real homes for life orbit. Although these homes will be our main focus of attention, the stars themselves are important for at least three reasons, one of which is very obvious, the other two less so.

The obvious reason stars are important to life is that they create the vast amounts of energy on which life depends – in the case of some life forms (photosynthesizers) directly, in others indirectly. A second reason for their importance is that their long lifespans provide sufficient time for evolution to occur on orbiting planets. This isn't true of all stars, as we'll shortly see, but it's true of most of them. The third reason is that life on Earth and other inhabited planets today would be impossible if it weren't for the existence of earlier stars. This is because, apart from trace quantities of lithium, no elements heavier than hydrogen and helium were present in the early universe; the heavier elements were all made inside stars. Let's now take these three aspects of the importance of stars in turn.

First, then, the creation of energy. A star is essentially a huge nuclear furnace. It's converting matter to energy by the process of nuclear fusion. Although what gets fused into what changes as a star begins to die, for the vast majority of its life it is fusing hydrogen into helium. Some mass is converted into energy in this process. Since the conversion operates according to Einstein's famous formula, each tiny unit of mass produces a vast amount of energy – because it gets multiplied by the speed of light squared, which is an enormous number. The amount of energy produced by a star per second is so great that it's hard to comprehend. Naturally, the precise amount depends on the size of the star. We can take the Sun as an example, as it's somewhere in the middle of the range of star sizes in terms of mass.

The Sun's power output is a massive 3.86×10^{26} watts (a watt is a unit of power which is defined as one unit of energy called a joule per second). Of course, only a small fraction of that reaches Earth, because we're just one tiny 'pale blue dot' (as Carl Sagan famously said) on a Sun-centred sphere of all the possible directions in which its energy can travel. But despite that, we still get 1.7×10^{17} watts. Although there's a further reduction because some solar energy is reflected back into space by clouds before it gets to the ground, radiation from the Sun powers almost the entire biological realm on Earth. There are exceptions – microbes that can get energy from inorganic compounds without using light, by a process called chemosynthesis. But in terms of the biosphere as a whole, they're a very small component. I'd guess that the same is true of most other inhabited planets.

Now we turn to the role of stars in providing a setting for orbiting planets that is stable in the long term – long enough for evolution to take place. That means stable for billions of years rather than merely millions. How long a star lives – and thus whether it can satisfy this criterion for life – depends on its mass, and the nature of this dependency is the opposite of what might at first be thought. On initial consideration, it might seem reasonable to expect more massive stars to live longer because they have more nuclear fuel in the first place. However, their higher temperature means that the fuel is used up more quickly, and this more than compensates for the fact that there was more of it when the star was born. So in fact massive stars are short-lived, middle-mass stars like our Sun live longer, and the most diminutive stars – red dwarfs – live longer still. Lifespan estimates are given in Figure 2.3 for the seven recognized star classes, each designated by a letter. Since the order of the letters was fixed historically rather than logically, it helps to have a mnemonic to remember it. A frequently used one is Oh Be A Fine Guy/Girl, Kiss Me! It would seem from these estimates that the most massive stars (categories O and B) don't live long enough for there to be life on their orbiting planets.

But here's an important caveat. I emphasized earlier that 'year' in a cosmic context is a variable rather than a constant. This is part of a more general phenomenon that many things that we're used to regarding as constants here on Earth become variables when we think more broadly across the universe. In the present context, the variable of interest is 'speed of evolution'. This is even more complex because it's a variable here on Earth. But imagine if we could

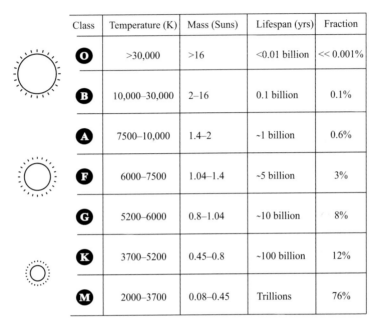

Class	Temperature (K)	Mass (Suns)	Lifespan (yrs)	Fraction
O	>30,000	>16	<0.01 billion	<< 0.001%
B	10,000–30,000	2–16	0.1 billion	0.1%
A	7500–10,000	1.4–2	~1 billion	0.6%
F	6000–7500	1.04–1.4	~5 billion	3%
G	5200–6000	0.8–1.04	~10 billion	8%
K	3700–5200	0.45–0.8	~100 billion	12%
M	2000–3700	0.08–0.45	Trillions	76%

Figure 2.3 The seven recognized classes of stars, together with various pieces of information on each, including their approximate lifespans. Also given are estimates of their relative commonness. As can be seen, the smallest stars – red dwarfs (class M) – endure for the longest time individually; and this class of stars collectively includes about three-quarters of the total. In contrast, the most massive stars (class O) live for a short time and are the least common. The Sun is a G-class star, with an estimated lifespan of about 10 billion years, which it's currently about halfway through.

estimate the *average* speed of evolution on Earth – might there be planets on which the equivalent of this figure is an order of magnitude higher or lower? It's hard to say, so we can't rule out the possibility of life on planets orbiting short-lived stars, but it seems improbable, especially given that planets suitable for life have to be very far out from these ultra-hot stars, and thus planetary orbits (and hence years) are longer rather than shorter.

It's time to turn to the third reason why stars are important for life – their manufacture of the elements of which life forms are made. Of these elements, only hydrogen existed in the early universe. Young and middle-aged stars, such as the Sun, make lots of helium, but our bodies have no need for this element. The elements heavier than helium are made in old stars, and especially in dying ones. The deaths of large stars in supernova explosions are particularly productive in this respect. Moreover, these explosions blast the heavy elements out into space. They permeate the clouds of gas and dust that are the birthplaces of later stars, with the result that planets orbiting these later stars have more of the materials for life than their predecessors. Indeed, the very earliest planets were probably not only all sterile but also all gaseous rather than solid, because the elements that rocks are made of didn't yet exist either.

Planetary Systems

Our conclusion so far is that stars are essential for life, but are not themselves habitable, at least as long as our definition of life includes metabolism. We now turn to the rocky bodies orbiting stars that *may* be homes to life. This takes us to planets – and perhaps also to moons and other bodies. At this point, it's necessary to have an in-depth understanding of the concept of a planetary system – which can be roughly defined as all the bodies, both rocky and gaseous, that orbit a central star – either directly, like planets and comets, or indirectly, like moons. In some cases, there are two or more central stars; binary and multiple-star systems are now known to be quite common.

For a long time, the only planetary system we knew of was our own – the solar system. This was discovered in stages. We've already noted the role of the ancient Greeks, in particular Aristarchus of Samos, plus the fact that his proposed heliocentric model didn't become widely accepted. But Copernicus's proposal, published in 1543 – the year of his death – in his book *On the Revolutions of the Celestial Spheres*, was gradually accepted, and indeed built on. The next major step was the working out of the laws of planetary motion by Johannes Kepler in the early seventeenth century. These laws included the recognition that orbits are ellipses rather than perfect circles. The discoveries of the outer planets were made in the eighteenth and nineteenth centuries (Uranus and Neptune, respectively). We've kept discovering things since then, of course, but the essential nature of the solar system as one with a single central Sun orbited by eight planets (and various smaller objects) has been known for more than 170 years.

In contrast, other planetary systems have only been discovered quite recently, starting in the 1990s. At the time of writing, about 3700 such systems are known. So far, we've discovered systems with every number of planets from one to eight, though we should bear in mind that our estimates are usually minimum numbers, because we might later discover more planets in any particular system. There are just two systems of eight planets known so far – our own, plus Kepler-90. And to date there have been no discoveries of systems with nine or more planets. However, with time, we should expect to see more cases of eight planets, some of nine and ten, and probably others with even more.

The discovery of lots of planetary systems has resulted in an appreciation of their diversity, but this is accompanied by an underlying uniformity – particularly in how they arise. A planetary system starts as a circumstellar disc – a collection of gases and dust orbiting a host proto-star. Over time, dust particles stick together, as do the resultant larger particles. This process keeps going through units of ever-larger mass. After a time, the disc consists largely of rocks of various sizes. Further accretion of these rocks results in bodies of more than a kilometre in width, which are called planetesimals. The process culminates with the formation of larger proto-planets and eventually planets. At the initial disc stage, particles stick together by electrostatic attraction. Later, gravity begins to play a role, as does a series of major collisions and mergers.

The form any one planetary system takes is determined much by chance, as argued by Stuart Ross Taylor in his 2012 book *Destiny or Chance Revisited*. Exactly what object collides with what other object is inherently unpredictable. Thus, some systems end up with more planets than others. Also, some end up with different types of planets. For example, some systems have 'hot Jupiters' – very large planets the size of Jupiter or even more massive – orbiting extremely close to their host star, closer than Mercury is to the Sun in our own system. The size and nature of any particular 'final' planet is much affected by the late-stage collisions in which it is involved. For example, it's thought that the Earth and the Moon ended up the way they did because of a collision between the proto-Earth and a smaller proto-planet called Theia.

In addition to eight planets, the solar system has several dwarf planets (including Pluto and Ceres), more than 150 moons (the exact number depends on the lower size limit used), and huge numbers of asteroids and comets. We think that all these types of object are routinely found in other planetary systems, even though they're harder to detect than planets due to their smaller sizes. There are now several known cases of exocomets, and some suspected exomoons. Others will doubtless follow in due course.

Which of these various kinds of orbiting bodies are likely to be hosts for life? The key to life – at least as we know it – seems to be liquid water. In our own system, we know of the occurrence of this on the surface of just one planet – Earth. We also know of subsurface water on several moons, such as Jupiter's Europa and Saturn's Enceladus. Regardless of whether life is eventually found in the subsurface oceans of these local moons (more on that in Chapter 5), life on moons seems likely to occur in at least some planetary systems, given their huge number. However, life on comets or asteroids seems much less likely, given their lack of either oceans or atmospheres. So I will generally ignore these from here on, and will focus on planets and moons.

Not all planets are to be found in planetary systems that orbit host stars. There are also planets that exist on their own in deep space. These are variously called rogue planets (the label I used earlier), lone planets, orphan planets or sunless planets. At first, they were thought to be very thinly distributed and very rare compared to 'ordinary' planets, but views on this have shifted over time, and rogue planets are now thought to be common – though exactly how common remains to be seen. There are two ways in which these planets could form. First, they may form *in situ*, in the same way as a star does – by the collapse of a cloud of gas and dust. Second, they may have originally been within a planetary system but were expelled from it. Either way, their lack of a sun makes life improbable. But perhaps not impossible – I've already mentioned the book *Imagined Life*, which argues the case for rogue-planet life. The basis of such life couldn't of course be photosynthesis, which depends on harnessing solar energy. Instead, it would have to be chemosynthesis, where other forms of energy are employed. Could chemosynthetic organisms arise from an evolutionary process on a rogue planet? Personally, I doubt it, but it's best to keep an open mind.

The Fourth Dimension: Time

This chapter has been mostly about space, as is appropriate when the question being asked is 'where to look for life?'. There isn't a counterpart question regarding time, in the form of 'when to look?', because we can only look in the present. However, complications arise in the temporal dimension for two reasons.

First, if we look far enough away in space, we're looking back in time too. If we look at a planet in the Andromeda galaxy, we're looking back about 2.5 million years. Imagine an Andromedan counterpart of Earth – a planet whose evolutionary tree is identical to that of our home planet. Of course, such a high degree of similarity is vanishingly unlikely, but this thought exercise serves a point. The light from Andromeda that we see through our telescopes today left there long before the first *Homo sapiens* evolved on the planet concerned. Given that our realistic search area is the Orion spur of the Milky Way, this complication isn't too important, because we're only looking back a few thousand years at most, rather than a few million.

Second, the exoplanets we look at as possible homes for life are many different ages. We've seen that stars have very varied lifespans. Thus, so too do their orbiting planets. These may be incinerated as their host star dies; if not, they could conceivably become rogue planets. Either way, those that were inhabited are likely to become sterile. The age of a planet is not just determined by its time of death; it is also determined by its time of birth. So a natural question becomes: when were the first planets born? The short answer to this is 'billions of years ago'. A longer answer requires a framework for universal time. I'll now give such a framework. It's not the usual one you'll find in textbooks on cosmology, though; instead, it's a framework from the perspective of life. So we won't look at the amazingly brief early timespans that interest cosmologists, such as the era of 'inflation', when the universe expanded dramatically – this was about 10^{-36} of a second after the Big Bang. Instead, we'll start with the first half-hour.

Phase 1: The First Half-Hour

From a few seconds after the Big Bang to about 15 minutes later was the era known as primordial nucleosynthesis. In general, nucleosynthesis is the process in which all elements get made from a starting point of the simplest one – hydrogen. Such synthesis comes in two forms, both of them dependent on extremely high temperatures. Primordial nucleosynthesis occurred pretty much everywhere in the early expanding universe. However, it ceased when the temperature dropped below the critical threshold. This is thought to have happened well before half an hour after the Big Bang. After that, there was no more nucleosynthesis until the first stars were born. Their appearance involved the second form of synthesis of helium and heavier elements – stellar nucleosynthesis, which is going on in our Sun and other stars right now. At the end of primordial nucleosynthesis, the temperature of the universe had dropped a lot since the Big Bang, but was still millions of degrees. Was there life back then? No, certainly not.

Phase 2: Up to About 400,000 Years

Primordial nucleosynthesis didn't produce atoms, just their nuclei. Nuclei didn't combine with electrons until the temperature of the expanding universe had dropped much further – to a mere thousands of degrees, rather than millions. This is thought to have happened at about 380,000 years after the Big Bang. Since there were only hydrogen and helium nuclei before this 'recombination' took place (plus trace quantities of lithium), the only atoms that were formed were of those elements. By the way, although it's usually called *re*combination, this choice of term is illogical, as it was the *first* such combination. Was there life when 'recombination' took place? Again, certainly not.

Phase 3: Up to About 100 Million Years

The era from when the universe was less than half a million years old until it was about 100 million years old can be referred to as the cosmic dark age. In this gap between one form of nucleosynthesis and another, no light was being produced. Things changed only when the first stars formed. The exact time of their formation isn't known, of course. But we think it was about 100 million years after the Big Bang, give or take quite a bit. These stars were probably patchily distributed, perhaps in proto-galaxies or smaller associations that later merged into galaxies. They may have had planets, but if so these must have been gaseous ones, because, as I mentioned earlier, the elements needed to make rocks weren't yet in existence. Since any such planets didn't have the elements needed for life either, there can have been no life at this stage in the universe's history.

Phase 4: Up to About Half a Billion Years

We've seen that some stars are very transient – notably the massive O and B class stars that live fast and die young. Their deaths in supernova explosions shower the space around them with many elements heavier than helium, including those needed to make rocky planets and life forms. When new stars form in these regions, life on their planets is possible. But how long does it take to evolve? If the speed of evolution there was about the same as on Earth, then the first proto-cells may have appeared after a few hundred million years. So we get a scenario in which the first stars are born when the universe is about 100 million years old, the first of these die after about another 100 million years, and the next stars in that area form after (say) another 100 million years. Assuming these stars had planetary systems, the first abodes for life came into being at this time too. But life takes time to establish itself through biochemical, and then early biological, evolution. If we guestimate that even the most rapid evolutionary system needs at least 200 million years to produce a biological result, then the first life forms could *conceivably* have arisen when the universe was 500 million – or half a billion – years old.

Phase 5: The Last 13.3 Billion Years

This vast span of time can be regarded as the age of possible life. So, for practical purposes, so long as we don't waste our energy searching for life on very young planets, everything else is fair game. This is especially true of the search for microbial life. For multicellular life, and especially intelligent life, we need to be a bit more restrictive, but even then we can look at planets that are anywhere between about 13 billion and 3 billion years old. Naturally, these are only to be found orbiting stars that are also at least this age, but there are plenty of those about. Stars of classes K and M born in the earliest waves of star formation are still around today, unless they've been destroyed in one-off cataclysmic events. And while the earliest G-class stars (the class our Sun is in) are long gone, those born anything up to about 10 billion years ago may still be in full swing. Life may exist on planets orbiting any of these stars. In some cases it may be old, in other cases young – but given the number of stars, you can bet it's 'out there' somewhere.

Two Perspectives on the Search

I hope that by this stage you feel that you've acquired a good general picture of the nature of the search for life in the universe beyond Earth, including the way it's restricted in space – and also in time. Here's a brief recap of the 'narrowing down' process, starting with the universe as a whole and ending up with our focus for study. First, we're inevitably restricted to the observable universe. Second, of its many galaxies, currently estimated at about a trillion, we focus on our own – the Milky Way. Third, because our galaxy is so large, we often focus on our local neighbourhood of it – the Orion spur. Within that spur, and hence within a few thousand light years (compared to the 100,000 light years diameter of the Milky Way as a whole) we focus the search on planetary systems. Other things being equal, the closer a system is, the better. Planets a few hundred light years away are easier to study than those a few thousand; and planets a few tens of light years distant are better still. Finally, we avoid looking at planets in very young systems, because life won't have had time to evolve there.

This focus applies to the search for life in general, via biosignatures such as oxygen, and also to the 'active' part of SETI, in other words the sending of messages intended to be read by alien civilizations. There's little point in sending messages to parts of the galaxy that won't arrive before all of the senders are dead. This is why the famous Arecibo message was really a test of procedures rather than a serious attempt to make contact – it was aimed at a star cluster that's 25,000 light years from Earth. However, the 'passive' part of SETI, searching for incoming radio signals sent towards us from alien civilizations, has a potentially wider scope, given that receipt of the message in itself would be an amazing event, regardless of how long it had spent in transit or our ability to reply within a meaningful time frame.

So, for most looking-for-life purposes, our search is fairly well circumscribed, for good practical reasons. It's focused on our own region of our own galaxy. However, this doesn't mean that it's only informative about that particular place. There are two ways of thinking about the investigation of our galactic neighbourhood. First, taking a narrow view and ignoring all other parts of the universe, our search of the Orion spur will eventually – we hope – produce clear evidence for life there. Second, taking a wider view, this search can be thought of as a sort of sampling exercise, whose ramifications transcend the Milky Way. What I'm getting at here is that sampling one arm of one spiral galaxy should be informative in a general way about any arm of any spiral galaxy. Thus, our 'local' search for life should tell us a lot – at least in the form of probabilities – about life in the universe as a whole.

3 Evolution – Here and Elsewhere

Controversy and Strategy

This is where we switch from the geography of alien life to its biology – in other words from its distribution across the observable universe to its 'nature' in many senses of that word, including its chemical composition, its physical form, its means of acquiring energy, and, in some cases, its intelligence. For me, the nature of life beyond Earth is even more interesting than exactly which planetary bodies it inhabits, and many other scientists feel likewise. However, in moving from geography to biology things also become more controversial, because the so-called 'sample size of one' problem comes into sharp focus.

Here it is, in a nutshell. To anticipate what life might be like on other planets, we inevitably tend to think in terms of 'life as we know it' here on Earth. And when considering the habitability of planets, there's usually a focus on their habitability for Earth-like life, as we'll see in the next chapter. But perhaps life elsewhere might be fundamentally different to life here, in ways that we simply can't imagine? Well, yes it might. However, a common view is that the probability of alien life being different in its *broad features* is low, while the probability of it being different in *detail* is high. This view leads to a strategy for how to deal with the issue. Here, I'll outline some of the important evolutionary steps that have taken place in the establishment of life's broad features on Earth. Collectively, these form a sort of case-study of the evolutionary system of an inhabited planet, which can be examined in its own right. As I discuss this case-study of evolution on Earth, I'll just make a few scattered comments on the extent to which there may be parallel features elsewhere. Later, in Chapter 7, I'll flip over the emphasis from Earth to other inhabited planets.

There, I'll focus on the question of the extent to which we might expect alien life to broadly resemble Earth life.

What I'm calling the *broad features* of life can be split into two classes: those that have persisted indefinitely through evolutionary time ever since the origin of life (next section); and those that have changed significantly as evolution has proceeded (subsequent ones). We could call these 'basics' and 'elaborations', or alternatively we could call them the 'time-independent' and 'time-dependent' classes of the broad features of Earthly life. For example, cellular structure is almost universal in Earth's living world, as we'll discuss below, and cells have existed since the beginning of life on Earth right up to the present day, so they can be classed as time-independent, or basic. In contrast, overall organismic form has varied much over time. Some variation in body form takes us beyond the realm of broad features and into detail – for example slight changes in body size or shape. However, some key aspects of body form qualify as broad features of the time-dependent kind. For example, the type of symmetry that characterizes an animal's body can be radial (e.g. jellyfish, sea anemone) or bilateral (e.g. butterfly, human). The evolution of bilateral from radial symmetry more than half a billion years ago was a particularly important event in the evolution of intelligence, as we'll see later in this chapter.

Origins of Life

Many discussions of the origin of life on Earth focus on the when and where of the matter, but I'm not going to follow that trend. Instead, I'm going to focus on the establishment of the basic, or time-independent, features of life on Earth. That's not to say that the place and time of life's origin are uninteresting – perish the thought. It's just that we understand them quite well in general terms but quite badly in specific ones. I'll now briefly state our understanding of the former (a single paragraph will suffice), and then move quickly on without wasting space on the unknown specifics.

So, here's our general understanding of the when and where issues. Earth is about 4.6 billion years old. Life began here about 4 billion years ago, give or take a wide margin of error. How wide? At least plus/minus a quarter of a billion years, and maybe even plus/minus half a billion – in other words, life arose sometime between 4.5 and 3.5 billion years ago. Also, we should always remember that 'origin of life' refers to a time-extended process, not a sudden, singular event. It was the process that led from unbounded collections of small and medium-sized organic molecules to rudimentary cells that contained these and also big ('macro') molecules, and were bounded by membranes. The process occurred in aquatic environments, but exactly which ones we don't yet know. One theory favours the ultra-hot water of deep oceanic hydrothermal vents, another the benign warm water of shallow seas. Might life have originated more than once on Earth? Yes, for sure. However, if it did, then the progeny of all but one origin died out. Although it is favoured by a few scientists, the idea that we'll eventually discover life forms on Earth that have a different origin to all those known to date, and are hence completely unrelated to them – so-called 'shadow life' – seems fanciful.

I'll now deal with three interrelated basics of life on Earth: carbon, water, and cells. Let's start with carbon. As I mentioned in the first chapter, some people argue that life could be based on a different element than carbon – silicon is the most commonly suggested, but there are others too. So why do I feel that those advocating silicon-based life are wrong to do so? The answer lies in a single word: information. All the activities of life forms require information. This is certainly true of Earth, and it's hard to see how things could be otherwise elsewhere. The everyday metabolic activities that sustain any life form on our home planet require highly specific reactions catalysed by highly specific enzymes – usually in the form of protein molecules. These are made by equally specific genes – usually in the form of DNA. The passing on of the genetic information necessary for offspring life forms to make their own enzymes and other proteins requires a highly specific mechanism – usually in the form of DNA replication.

Given that life elsewhere must involve flux of the complex information without which neither metabolism nor reproduction-with-inheritance could operate, the question becomes one of whether non-organic (i.e. non-carbon-based) molecules exist – or could be made – that could rival the information-storage capabilities of DNA. So far, the answer seems to be that they do not exist – either on Earth or anywhere else in the universe. That's not to say that other elements can't be the basis of large molecules – they can. For example, silica is a big molecule involving a mixture of silicon and oxygen. But its structure is much simpler than that of DNA, and consequently its informational capacity is much lower.

Does this line of reasoning lead to the conclusion not only that life must be carbon-based but also that it must be DNA-based? The answer is 'no' even here on Earth. While there are no Earthly life forms without carbon, there are some that don't use DNA as their genetic material. For most of the course of evolution, the only exceptions have been viruses, and we've seen that it's debatable whether these should be described as being truly alive. But at the very earliest stages of evolution, life on Earth may have gone through a phase in which all life forms used RNA instead of DNA. This is the 'RNA world hypothesis'; if it's correct, then non-DNA-based life was once common on Earth. Perhaps there are planets on which the RNA world lasted indefinitely. Perhaps there are even planets on which the genetic material is made of proteins. That seems unlikely to me, but we can't rule it out.

Now to water. From our aquatic origin of life in the distant past to all the inhabitants of today's biosphere, Earthly organisms have been dependent on water. The exceptions are so few and so odd that they effectively 'prove the rule'. For example, an anhydrous tardigrade tun is a life form (refer back to Figure 1.2), but not a functioning one. What form does our dependence on water take? The main role of water in life on Earth is to act as a solvent. The importance of this is that metabolic reactions generally happen in solution. Water has, in fact, been described as the 'universal solvent', but this is a misleading term. It would perhaps be better called the optimal solvent instead, because the key point is not that water dissolves everything (it doesn't), but rather that it dissolves more compounds than does any other known solvent. The substances it dissolves range from simple salts to organic macromolecules, including many proteins.

The fact that water doesn't dissolve all substances is as important as the fact that it dissolves many of them. Water-insoluble substances play a key role in separating the aqueous cytoplasm of organisms into manageable units (cells), distinct from the outside environment and – in the case of multicellular organisms – from each other. Water can't dissolve lipids (fats); these are a major component of cell membranes. This is true throughout the living world on Earth. I'd guess it's true of alien life also, but that's not a given.

This discussion of cell membranes leads naturally to a broader discussion of cells, the final one of our three basics of Earth life. Cells aren't quite as ubiquitous among Earthly life forms as the other two basics – carbon and water – but they come close. Out of trillions of organisms belonging to millions of species on our home planet, there is not a single exception to the rule of being carbon-based and – while metabolizing – water-based. Is there a single exception to being based on cells? The answer depends on exactly how the question is phrased. Here are two specific versions that lead to different answers.

First: Is there any kind of 'true' life form on Earth that does not take the form of one or more cells at *any stage* in its life cycle? If we interpret 'true' to mean 'satisfying all three of the criteria of the RIM definition of life' – thus excluding viruses – the answer is 'no'. With this way of phrasing it, then, the situation with regard to being cell-based is just as clear as it is with being carbon-based or water-based – there no exceptions. Of course, if we skip the 'true' and count viruses in, then the answer changes accordingly.

Second: Is there any kind of 'true' life form on Earth that does not take the form of one or more cells at *some particular stage* in its life cycle? Now the answer is 'yes'. The main exception to cellularity when the question is phrased in this way is the group of organisms known as the acellular slime moulds. These have a cellular phase in their life cycle associated with reproduction, but for most of the lifespan they consist of a large and rather amorphous structure that's not divided into cells, called a syncytium (meaning 'fused cells'), as shown in Figure 3.1. This group of slime moulds consists of only about 1000 known species, out of a total of millions of species of life on Earth overall, so it's a small exception. Syncytia also appear in certain tissues or developmental stages of many other organisms, including part of the mammalian placenta and an early stage in insect development called the syncytial blastoderm; but such appearances are against a background of cellularity more generally.

Our conclusion, then, is that the single (or single successful) origin of life on Earth produced three basic features of life that are ubiquitous (carbon, water), or very nearly so (cells), among descendant organisms right up to the present day, and perhaps until life on Earth is extinguished by the increasing luminosity of the Sun, a few billion years into the future. I suspect that these three features will characterize most or all alien life too.

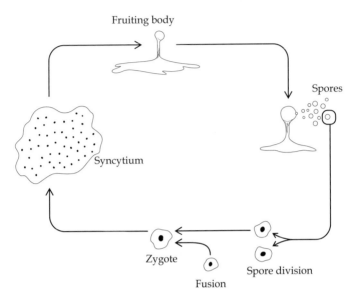

Figure 3.1   Non-cellular life on Earth? The life cycle of an 'acellular' slime mould (a member of the group called Myxogastria) shows that 'acellular' is a simplification in that it refers to the adult form but not the reproductive stages. Sexual reproduction involves the fusion of two individual cells, which are haploid and of different 'mating types', producing a diploid zygote, again an individual cell. As the zygote nucleus begins to proliferate, this process is not accompanied by cell division. This means that we end up not with a large number of cells but rather a large number of nuclei – up to millions of them – in a continuous 'sea' of cytoplasm. This structure is called a syncytium. Eventually, fruiting bodies arise from the syncytium. In these, meiosis occurs and haploid reproductive cells (spores) are produced, which begin to divide, thus completing the cycle.

Proliferating Microbes

Now we begin to follow the course that evolution on Earth took after the end of the 'origin of life' phase, focusing on a small number of key events. Let's start by trying to picture the four-billion-year old Earth. It's not easy, because we have little direct evidence of what it looked like, but here are a few guidelines.

There were oceans. There were also emergent landmasses. The land wasn't organized as it is now, into our familiar continents, but that's not important. The Earth was a bit cooler than it is today, because back then the Sun was less luminous than its present-day self. Despite this fact, some water evaporated from the oceans. This ultimately precipitated back to Earth as rain and snow. Thus, there was a water cycle. There were also rivers and lakes as parts of this cyclical flow of the all-important liquid.

This is the environmental background against which evolutionary lineages of the very first single-celled life forms evolved their way into the future. And for a long time, all of them remained single-celled. The first known multicellular organisms – red seaweeds – didn't grace the Earth with their presence until almost three billion years later. But let's pause here for a moment to consider the apparently neat binary division of life forms into simple unicellular and complex multicellular life. As is usual in biology, apparently neat divisions aren't neat at all. There are many possible kinds of construction that fall in between the two. They include filaments, films, and mats of cells, as well as more irregular loose associations that could collectively be described as colonies.

There's a different division of life forms on Earth that comes much closer to being a perfect binary split than single-celled versus multi-celled. This is the split between *prokaryotic* and *eukaryotic*. If you're familiar with this, then please skip to the next paragraph. The cells of a human are eukaryotic (which roughly translates as true-kernel, in the sense of having a clearly defined nucleus). Their central nucleus is surrounded by cytoplasm, in which are located various organelles, for example the energy-releasing ones called mitochondria. The cytoplasm is bounded by a two-layered lipid-based membrane, as are the nucleus and the organelles. A bacterial cell is of the other kind – prokaryotic (which roughly translates as lacking a kernel). Here, the genetic material just exists as an unbound loose grouping of DNA and other molecules, often in the central area of the cell, and sometimes referred to as a nucleoid. There are no membrane-bound organelles. The only membrane is that which surrounds the cell itself. Clearly, prokaryotic cells are simpler than their eukaryotic counterparts, though it's important to understand that this is a comparative description; no cells merit the unadorned label 'simple'.

Now let's connect the relatively clean division between prokaryotic and eukaryotic cells with the messier one between single-celled and multi-celled body construction. Bacteria and the other group of organisms based on prokaryotic cells, the less-well-known Archaea, are either unicellular or what we might call minimally multicellular – forming the filaments, mats, and irregular 'colonies' referred to above. Occasionally, they form structures that are more impressive than these. An example is provided by the fruiting bodies of a group called the Myxobacteria (Figure 3.2). These have multiple cells with a clear three-dimensional form that includes structures that could be referred to as stalks and spheres. However, even this pinnacle of bacterial complexity lacks any equivalent of the organs that we find in the multicellular bodies of eukaryotic organisms such as humans (e.g. brains) or orchids (e.g. flowers). Nonetheless, prokaryotic organisms can be multicellular, albeit minimally so. Now we also need to acknowledge that eukaryotic organisms can be unicellular – for example, yeasts, many forms of red and green algae, plus other tiny creatures such as amoebae (singular amoeba) and a large group of parasitic forms that includes the malarial parasite *Plasmodium*.

At this point I should explain my use of 'microbes', which is essentially a short version of 'microorganisms'. I include all of the following under this

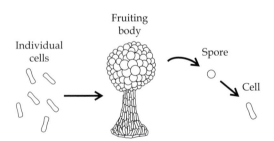

Figure 3.2 Multicellular fruiting body of a member of the group Myxobacteria, releasing spores, which 'germinate' into individual cells. These bodies can contain hundreds of cells, and, as can be seen, have a definite overall structure consisting of a stalk leading to spherical collections of spores. These multicellular structures defy the usual notion of bacteria as organisms that are unicellular throughout their life cycles. It's true that some bacteria *are* always unicellular, but clearly the Myxobacteria are not.

umbrella-term: all prokaryotes, whether unicellular or minimally multicellular; all unicellular eukaryotes; all eukaryotes whose life cycles culminate in forms that also fall under the 'minimally multicellular' heading, including hollow balls of cells and other such structures. In contrast, organisms with the kind of multicellularity that includes tissues and organs are not microbes, however small they may be – recall those tiny tardigrades, many of which are much less than a millimetre long.

Using this terminology, the first three billion years of evolution on Earth was an entirely microbial affair. Parental lineages of microbes diverged into descendant lineages; the descendant ones gave rise to their own descendants, and so on for countless generations. In this way, many different types of microbes were produced. The driving force was, of course, Darwinian natural selection. Evolution worked the same way back then as it does now. Populations adapt to their environments. Some of the early microbes were extremophiles, tolerant of environments that were very hot, very cold, very acidic, very alkaline, or subject to other extreme conditions. Other microbes lived in relatively benign environments. And different early microbes acquired their energy in different ways, just as do different microbes alive today.

As microbial lineages diverged from each other, the number of them that coexisted on Earth gradually grew, despite the fact that from time to time some of them went extinct. This number, which must have started as one, became tens, then hundreds, then thousands. Most of the divergences in the microbial tree of life produced descendant organisms that weren't very different from their progenitors. But occasionally something crucial happened. So what we need to do is to ignore all the minor-effect splits and focus instead on those few cases in which some key innovation took place.

In my view, there were three particularly important splits. First, the split between bacteria and archaea. Although most archaeal cells look superficially quite like bacterial ones, there are some deep-seated differences. For example, the cell membranes and cell walls have some differences in chemical composition, the genes are different in structure, and some parts of metabolism follow different pathways.

The second key split occurred within the bacteria. Branching off some already-established bacterial lineage was a new one that led to the cyanobacteria (previously called the blue-green algae). These organisms have an importance for the biosphere that transcends their minuscule size. They are the only group of bacteria – indeed the only group of organisms based on prokaryotic cells – that obtain their energy from sunlight via photosynthesis and produce oxygen as a by-product. As the cyanobacterial lineage split and split again, eventually producing thousands of species, and as these spread across the globe and all went about their business of photosynthesising, the composition of the Earth's atmosphere began to change. Starting about 2.5 billion years ago (BYA), oxygen began to accumulate, from a starting point of virtually zero. Between 2 and 1 billion years ago it plateaued at a few percent, then in the last billion years it increased further, eventually reaching its current value of about 20%. The initial increase, between about 2.5 and 2 BYA, is referred to as the Great Oxygenation Event. This event changed everything for life. It caused the extinction of many organisms to which oxygen was a poison, but at the same time opened up a whole new raft of possibilities for the evolution of organisms with aerobic (oxygen-using) metabolism.

The third key split in the early microbial evolutionary tree was the one that produced the first organisms based on eukaryotic cells – organisms that would turn out to be the ancestors of all animals, plants, and fungi. What happened here was that one type of prokaryotic cell became embedded within another, and the composite organism thus formed evolved to become a single functioning entity. Actually, this seems to have happened more than once. On one occasion, a small cell, probably from the bacterial group Rickettsiales, became embedded in a large cell, probably an archaeal one, and evolved into a mitochondrion – a process known as endosymbiosis. This provides the basis for the unicellular ancestors of animals and fungi. Later, some descendant of the first endosymbiotic event took in another small cell, almost certainly belonging to the cyanobacteria, and this evolved into a chloroplast, thus providing a lineage that would lead to plants. Later again, yet another endosymbiotic event led to the origin of a photosynthesizing group that is quite distinct from the plant kingdom – the group that includes the unicellular diatoms and the multicellular brown algae.

Can we put a timescale to these key splits in the evolution of microbes? Well, yes, but only in very approximate terms. The split between bacteria and archaea happened sometime between 4 and 3 BYA. Cyanobacteria probably split from other bacterial groups between 3 and 2.5 BYA. Eukaryotes are thought to have arisen between 2 and 1.5 BYA. And the oldest known fossils of multicellular eukaryotes – the red algae called *Bangiomorpha* – date to about 1 BYA. At the 1 BYA marker, there were also green algae, but possibly only unicellular ones. There were no land plants at all. A similar situation may have prevailed in the fungal kingdom, with some unicellular fungi but no multicellular ones. And there weren't any animals at all. These days, multicellularity is part of the definition of an animal. Members of groups of unicells that would later give rise to the animals were in existence, but not animals themselves.

In conclusion, in some ways the biota of planet Earth about a billion years ago wasn't much different to its counterpart three billion years earlier – it consisted of a variety of microbes, with large complex organisms nowhere to be seen. But in other ways it was significantly altered. Some of the microbes present a billion years ago photosynthesized; these and their ancestors had transformed Earth's atmosphere, and in so doing had altered the course of subsequent evolution. Others were based on a type of cell that hadn't existed when life began, a cell-type that would prove to be a much better building block for the evolution of large multicellular creatures. Perhaps the sorts of evolutionary processes involved in the evolution of this building block – including endosymbiosis – have also played a role in the evolution of life on other planets.

Evolution of Multicellularity

One thing we know for certain about evolution on Earth is that every multicellular creature on the planet – including humans – had unicellular ancestors. So there must have been at least one origin of the multicellular condition. But was there only one, or were there more than that? And if there were more, how many? Answering this question, and many others like it, requires a good understanding of what's sometimes called 'the pattern of natural classification' – in other words, the structure of the tree of life. Luckily, while our understanding of the tree's structure was very imperfect at first – with fungi considered to be plants, for example – it is now much better, though of course still imperfect. Much of the recent progress in this area has been due to the use of molecular data – primarily DNA sequences.

The Swedish naturalist Carl Linnaeus made a heroic attempt to classify the living world in the mid-eighteenth century. Several other major figures should be mentioned. One was the French naturalist Georges Cuvier, who proposed, in the early nineteenth century, a structure for the animal kingdom involving four major divisions. Another was the American botanist Robert Whittaker, who proposed the five-kingdom tree of life in 1969. And in terms of *methods* of determining the structure of the tree of life, the German taxonomist Willi Hennig was a leading figure – he devised an explicitly lineage-based system (the one now called cladistics).

Here are some things we've learned at various stages between Linnaeus's time and the present. Not only are fungi not plants, nor are brown seaweeds such as the various species of kelp. Although they photosynthesize, their chloroplasts result from a different endosymbiotic event than the one described in the previous section that led to the plant kingdom. There is no animal grouping of 'radiates' (as Cuvier had proposed) – animals with radial symmetry; rather, starfish and their kin (echinoderms) had bilaterally symmetrical ancestors and have returned to a radially symmetrical adult body plan, while jellyfish and their kin (Cnidaria) were radially symmetrical from the start. In the animal tree of life they are only the most distant of relatives. There is no animal supergroup of 'segmented invertebrates' incorporating the segmented worms (annelids, such as the familiar earthworm) and the arthropods (insects and their kin); instead there are supergroups based on the presence or absence of growth-by-moulting, and a spiral pattern of cell division in early embryogenesis. The potato blight that caused the great Irish famine of the nineteenth century is not a fungus; rather it's a member of a kingdom that wasn't even on our radar in the 1960s, when Whittaker published his five-kingdom model. Despite its lack of photosynthetic ability, it's a member of the same kingdom as the brown algae. This kingdom is often now referred to as the heterokonts, with the *hetero* (different) in its title referring to the different lengths of two hair-like projections from the cells of these creatures, which is one of their main common features.

Why is having a good knowledge of the broad structure of the tree of life important for dealing with the question of how many times multicellularity has originated on Earth? Perhaps you feel that the answer is obvious, perhaps not. Anyhow, here it is, as a mixture of verbal and pictorial information. In Figure 3.3, I show two possible trees of life, one with a single origin of multicellularity (definitely wrong), the other with multiple origins (correct in broad terms, though its four origins are certainly an underestimate of the true number). What I had in mind when drawing the lower picture of Figure 3.3 were the four kingdoms of animals, plants, fungi, and heterokonts. But Animalia may be the only one of these kingdoms in which multicellularity has arisen just once. When you look in more detail at the structure of the other kingdoms, things get messier. Multicellularity has arisen at least twice in the fungi, and at least twice in plants. A complicating factor with plants is the fact that how you define the plant kingdom these days is debatable. There are three possibilities: just the land plants; land plants plus green algae; that

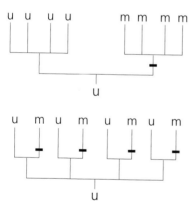

Figure 3.3 Two scenarios for the evolutionary origin of multicellularity on Earth. Top: a single-origin model, which has long been known to be incorrect. Bottom: a multiple-origin model, showing four separate origins in four separate kingdoms. This is closer to the truth but is nevertheless a simplification. Some of the eukaryotic kingdoms have experienced two or more origins of multicellularity, and at least one of them – fungi – has experienced a reversion from multicellular to unicellular forms (yeasts). Also, as we've seen, some prokaryotes have multicellular stages in their life cycles. u, unicellular forms; m, multicellular forms; horizontal bars, evolutionary changes from one to the other.

combination plus the red algae. So, following this line of reasoning, and looking in more detail at the kingdoms concerned, the number of origins of multicellularity rises from four to at least six.

However, it's easy to find claims in the literature that the true number is actually more like 20 or even 30, rather than a mere four, five, or six. The main reason for this difference is the adoption of more versus less inclusive definitions of multicellularity. Figure 3.3 only incorporates origins of eukaryote multicellularity; it's based on the view that the number of origins of 'true' multicellularity in the prokaryotes is zero. But recall that this is a debatable point of view, given the existence of a degree of multicellularity in the life cycles of Myxobacteria (refer back to Figure 3.2) and some other prokaryotic groups.

Any particular origin of multicellularity is just a beginning for the lineage concerned. A much bigger issue is what happens next – in the sense of the next millions or even billions of years. We're now approaching the fascinating subject of evolutionary trends in the complexity of organisms. This is the proverbial can of worms, but I'll try to steer a safe course through the difficulties by highlighting three key generalizations, as follows.

First, despite the laments of some pessimists, the complexity of an organism *can* be defined. One good definition is the number of different cell-types of which it is composed. Another is its number of different types of organs. Luckily, these two definitions produce broadly parallel scales of complexity, because the more organs an organism has, the more cell-types it must have too, as a result of the fact that different organs (e.g. brain, heart) are to a large degree composed of different types of cells.

Second, there is no 'law of increasing complexity over time'. Many evolutionary lineages have consisted entirely of unicellular organisms from billions of years ago to the present day, thus representing a 'flat-lining' of complexity. This is almost certainly the commonest pattern. It can happen right from the start, as in the unicell example, or after a certain degree of complexity has been reached. Birds provide an example of the latter situation. In general, birds are quite complex – especially when compared with unicells. But has the evolution of about 10,000 species of birds from a single ancestral species produced a significant increase (or decrease) in complexity? I think not.

Increases in organismic complexity over time occur in *some* lineages of organisms for sure, but by no means all of them over all time periods. Not only that, but in some lineages complexity can decrease over time. An example of this is the lineage that led to today's unicellular yeasts from multicellular ancestors.

Third, despite the fact that any particular lineage can produce increased, decreased, or flat-lining complexity over time, an overall trend in the evolution of complexity can be discerned if we focus on how *average* organismic complexity behaves as evolution proceeds. This is shown in Figure 3.4, in the form of average complexity rising but then tending

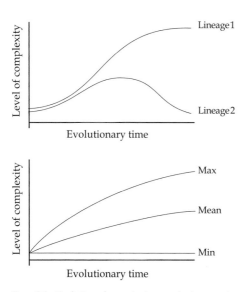

Figure 3.4   Evolution of organismic complexity over time. Top: two possible patterns in individual lineages, one showing an increase throughout the period concerned, the other showing an increase followed by a decrease. Often, an individual lineage will show no change in complexity over a shortish period, such as a few million years. Bottom: How the minimum, maximum, and mean organismic complexity have changed in the long term. Note that the minimum has stayed put – unicellular forms are still abundant today – while both the mean and the maximum have increased.

towards a plateau. This illustration is just a simplified model of what has actually happened in evolution to date; and given that what *will* happen over the next few billion years is unknown, the plateau may be illusory. But the increasing average to date is not. It has certainly been more erratic, and perhaps more stepwise, than the gradual ascent illustrated, but the increase is without doubt, as long as we consider long-enough periods of time.

Exactly what form the long-term increase in average organismic complexity takes, and the nature of the bodies it produces, naturally depends on which branch of the tree of life we consider. The trees of African rainforests are more complex by far than their green algal ancestors, and the apes that inhabit those forests are more complex by far than the small marine flatworms from which they ultimately arose. But the complex body forms of trees and apes have little in common, apart from the 'basics' that derive from the origin of life and *very* early evolution (notably cells) and the mere fact that they are indeed both multicellular. Also, *within* any one kingdom, there are high points of complexity that are very different from each other – for example, arthropods, vertebrates, and molluscs in the animal kingdom.

Clearly, in a book of this kind, there isn't space for an examination of all the different complex body forms that have arisen in the course of evolution – but that's no problem because there are many good books on this subject. Here, I'll restrict the discussion to just two increases in complexity – those associated with the evolution of land plants (next section) and the evolution of intelligent animals (the following one). There's a good reason behind these choices, and it's to do with the search for life in the universe beyond Earth. Photosynthesis by plants (as well as by cyanobacteria and some heterokonts) can alter planetary atmospheres, which leads to a possible means of detection of extraterrestrial equivalents of our own photosynthesizing organisms. Intelligent life can also modify planetary atmospheres (though not usually for the better), again leading to possible methods of detection. The gases being detected are *biosignatures* in the first case, *technosignatures* in the second. And of course intelligent life forms can also advertise their presence with other kinds of technosignature – radio and laser signals.

Evolution of Complex Plants

Let's take a quick look at the evolution of organismic complexity in the plant kingdom, before moving on to animals and their intelligence. There's an interesting high-level difference between plants and animals in terms of the connection between complex body forms and types of environment: the most complex animals are split between terrestrial and aquatic habitats, while *all* of the most complex plants are terrestrial. The most complex animal body plan is – arguably – the vertebrate one, which began in the seas and is now almost equally represented on land and in water, in terms of numbers of species. In contrast, the most complex plant body plan – that of the angiosperms (flowering plants) – began on land and has remained on land, albeit some wetland forms could perhaps be described as semi-aquatic. The plants of the oceans remain simple in form – almost all of them are algae.

Several key innovations were involved in the evolution of land plants. One was the evolution of systems of transport for both water and nutrients around the body. These tubular systems – xylem for water and phloem for the products of photosynthesis – originated at a certain point in plant evolution, namely the point at which the stem lineage leading to all of today's vascular plants (including ferns, conifers, and angiosperms) branched off from a lineage of moss-like plants. Living mosses and their kin (bryophytes) remain non-vascular.

Another key innovation was the transition from unicellular spores to multicellular seeds in the reproductive system. Although ferns are more complex than mosses in that they have a vascular system, they have spores rather than seeds. Seeds originated at a later split – the one at which the lineage leading to both conifers and angiosperms branched off from a lineage of fern-like plants. As the lineage leading only to angiosperms branched off from other seed plants, flowers appeared.

The final plant evolutionary innovation I want to mention is the origin of trees. However, this is more like the origin of multicellularity than the origin of flowers, in that it happened multiple times. Exactly how many depends, as you might expect, on how 'tree' is defined. The majority of tree species found today are angiosperms, but they don't all derive from a single angiosperm stem lineage; rather, many such lineages have evolved into trees independently of each other, as well as independently of the origins of coniferous trees and tree-ferns. In all cases where trees have evolved from non-woody plants, natural selection for increased reception of sunlight by over-growing competing plants is likely to have been involved.

In what way are these key innovations in the evolution of plant body form important, when considering the possibility of life beyond Earth? Well, in the long term, we'll be fascinated to find out whether parallel processes happened in the plant kingdoms of other planets. But in the short term, while we're still seeking evidence of *any* extraterrestrial life, knowing about these innovations doesn't seem to help much. What I mean by this is that although detection of a significant oxygen signal in an exoplanetary atmosphere would be suggestive of photosynthetic life, and thus hugely exciting, it wouldn't tell us whether the organisms producing the oxygen were alien versions of cyanobacteria or rainforest trees.

But perhaps there's a final twist to the story. The strength of an oxygen signal would depend on several factors, one of which would be the actual concentration of oxygen in the atmosphere of the planet concerned. Perhaps a densely forested planet would tend to give rise to a higher level of oxygen than one with just a thin gruel of photosynthetic bacteria suspended in nutrient-poor waters, other things being equal. Earth's atmosphere has a higher level of oxygen now than it did after the Great Oxygenation Event two billion or so years ago. We'll return to the important issue of detecting oxygen in exoplanetary atmospheres in Chapter 6.

Evolution of Intelligent Animals

I'm now going to look at a series of key innovations in the evolution of animals. In some ways, this story runs parallel the previous one on key innovations in plants. However, it has a different relationship with the issue of detecting life on other planets. In the case of animals, the innovations we'll look at are those that *led to* the kind of intelligence that can produce detectable technosignatures. In the case of plants, the innovations we looked at didn't lead to the process of photosynthesis – that had originated much earlier with cyanobacteria – but rather entailed major elaboration of the photosynthetic machines that we call plants from simple early forms to complex present-day ones. These plant innovations have increased the overall rate of photosynthesis on our planet, but they didn't create it in the first place.

In the introductory chapter, I said two main things about intelligence: first, that its evolution in the animal kingdom was a gradual process; second, that for the purpose of searching for intelligent life elsewhere, the pragmatic approach was to adopt a definition that connects with our search methods. Since radio signals play such a key role, using radio-level capability as our definition seemed a good choice, with the proviso that the level of brain complexity required to understand the production and receipt of radio signals evolved significantly earlier than the appearance of the first radio technology. Has a chimp the mental ability to understand radio? No. If a newborn human from 1000 years ago had somehow been cryo-preserved and then revitalized and brought up in the twenty-first century, would he or she have had this mental ability? Almost certainly yes. What about intermediate scenarios such as Neanderthals? Probably not, but we don't know for sure.

What key innovations were involved in the evolution of animal intelligence from zero to a human level? There are at least six of them, following the origin of multicellularity that was involved in the origin of animals themselves: (1) movement, powered by muscles whose contractions are coordinated by a nervous system; (2) bilateral symmetry; (3) cephalization; (4) manipulative appendages; (5) sociality; and (6) complex language. Let's consider them in that order.

First, then, movement/muscles/nerves. As far as we know, the earliest animals didn't move. We're not sure exactly what they were, but probably primitive sponges. Today's sponges still lack muscles and nerves. These tissues first arose in radially symmetrical animals that branched off the sponge lineage. Today's jellyfish exemplify that kind of body layout. They have a diffuse nerve net spanning the body, but no brain. The process leading to a large brain – cephalization – couldn't even get started in a radially symmetrical creature, because there is no head. Instead of having a head-to-tail axis, a jellyfish just has an oral–aboral (or mouth versus non-mouth) axis.

Second, bilateral symmetry. In one early lineage of animals, evolution led to a major change in body form, producing this kind of symmetry of the overall body for the first time. A bilaterally symmetrical animal – whether a human, a fish, or a butterfly – has three body axes rather than just one. The primary axis – usually the most elongate – is head-to-tail. The others are dorsal–ventral (back-to-front) and left–right. The evolution of an animal with a head end acts to channel further evolution in important ways. In particular, because it is associated with directional movement – there is now a 'forward' and 'reverse' – it leads to selection for a concentration of sense organs and nerve cells at the head end of the animal, which is the bit that first enters the unknown of a new place. This is the start of brain evolution, or cephalization. If you were to ask me what event in the entire history of the animal kingdom was the most important for the evolution of intelligence, I'd say without hesitation the origin of bilateral symmetry, even though the resultant creatures – tiny marine worms something like today's acoel flatworms – were barely above zero on any scale of intelligence that might be used for the animal kingdom as a whole.

Third, cephalization. From the minuscule concentrations of nerve cells at the head ends of early bilaterians arose brains of various sizes in various lineages. Some remained small, some got a bit bigger, and some ended up huge. In general terms, the most cephalized animal phyla are the vertebrates, the molluscs, and the arthropods. Within these, as I mentioned in Chapter 1, there are particular peaks of brain structure and intelligence in the mammals and cephalopods, though there's not such a clear a peak within the arthropods. Unlike the origin of bilateral symmetry, which was a one-off event, cephalization has been a protracted and gradual process, with plenty of lineages not showing it, and some even showing the opposite – which could be called de-cephalization.

Fourth, the evolution of manipulative appendages. This has been associated with the evolution of particularly high levels of intelligence in the apes and the octopuses. It isn't a prerequisite for high intelligence, as shown by dolphins. But it certainly helps. We know in some detail how human hands evolved, against a background of primates becoming arboreal and then proto-human apes descending from the trees, freeing up their hands for the manufacture and use of tools. It's less clear how or why the arms and suckers of cephalopods evolved from something akin to the (non-manipulative) foot of a snail. But in both cases, the evolution of manipulative appendages was associated with the evolution of high intelligence. Like cephalization, appendage modification has been a slow process, but not a regular one, in that there have clearly been bursts of it interspersed with periods of lower rates of evolutionary change.

Fifth, the evolution of sociality. Many kinds of animal live solitary lives, coming together only occasionally, for example to mate. Others live in dense associations, but with little behavioural interaction, for example the tightly packed groups of tube worms that live in the vicinity of hydrothermal vents at the bottom of oceans. There are also colonial forms, where there is more interaction, but the individual in a way loses its identity. Examples of colonial animals include corals, moss animals (bryozoans), and some sea squirts (tunicates, close relatives of the vertebrates). But what seems to be associated with the evolution of intelligence is something else again – social organization that falls short of coloniality, and involves enhancement rather than eclipse of the individual. A few octopuses show this, but strangely most are towards the solitary end of the spectrum. This kind of social organization is particularly common in mammals and birds. Picture, for example, a rookery, a school of dolphins, a troop of chimpanzees hunting monkeys, or a human city. Sociality has evolved many times, with very different results. Contrast, for example, the societies of humans and bees.

Finally, the evolution of language. Vocal communication is common in birds and mammals. Some of it can be quite complex – for example the various constituent strands of the avian 'dawn chorus' or the vocalizations of elephants, to which labels like 'bellowing' don't do justice. Whales and dolphins also have complex vocalization, some of it at a frequency too high for humans to hear. Their use of sound is informally referred to as whale-song. But in terms of language complexity, we humans have reached a point that is virtually 'off the scale' from the perspective of the animal kingdom as a whole. The evolutionary elaborations of our language and our intelligence have been deeply intertwined.

In conclusion, then, several key innovations have been important in the evolution of high intelligence in certain evolutionary lineages of animals, including the one that led to humans. But we shouldn't consider these innovations in isolation from each other. For example, bilateral body form made cephalization possible, cephalization made complex behaviour possible, and manipulative arms aided that process. Each innovation has built on those that went before it. And of course all of them were built on the origin of the nervous system in an ancient animal that's lost in the proverbial mists of time.

4 The Key Concept of Habitability

Habitable by What?

People involved in the modern era of the search for life – from the early days of SETI in the 1960s to the present – have tended to think of some planets as being potentially 'habitable' (or 'inhabitable', though that synonym is rarely used) and others not. For example, in our own system, Mars might once have been habitable, but Jupiter never so. Why this apparent certainty about Jupiter's unsuitability for life? There are two main reasons.

First, as a gas giant, it has neither a solid surface for life forms to live *on* nor bodies of liquid for them to live *in*. We think of the habitats for life as requiring a physical basis that isn't entirely gaseous. There's a complication here, because the phrase 'gas giant' may be misleading. Jupiter may have a solid core consisting of a mixture of rock, ice, and metals. But if it does have such a core, it's sufficiently buried below layers of swirling gases that no light reaches it.

Second, because of its outer location in the system, Jupiter is very cold. Exactly how cold is hard to say, as we normally think of a planet's temperature as that prevailing at its surface. Even that may be very variable, as comparison of Earth's polar and equatorial regions shows; but if we consider the up–down dimension as well as the pole-to-equator one, Earth's temperature variation is even greater. The core is thousands of degrees Celsius, while some layers of the atmosphere are permanently sub-zero. The *average* surface temperature of our planet is about 15 degrees Celsius. The best comparison we can make for Jupiter is its average temperature at the level in its atmosphere where the pressure is the same as at the surface of the Earth. This is about minus 150 degrees Celsius. Again, though, there are complications. Was Jupiter always at its present location? Probably not. Orbits in the solar system are relatively stable now, but early in the system's history orbital migration was common. However, the planet has been in the outer reaches of the system for long enough that it has been icy for most of its life to date.

So, although there are lots of ifs and buts, Jupiter is made of gas and is very cold. We interpret these two facts as meaning that it's not habitable to life. If by life we mean human life, that's certainly true. If we broaden out to 'any Earthly life form' then it's probably still true; even the most bizarre extremophile microbes we've discovered so far on Earth would find survival on Jupiter impossible. However, let's broaden out further again to life in general, according to the RIM definition of Chapter 1. Might there be some form of metabolizing life capable of reproduction-with-inheritance for which Jupiter might be habitable? For any such life that's based on carbon, water, and cells, the answer is almost certainly 'no'. But if we discard that restriction and imagine some completely different form of metabolism, then it's hard to say.

Naturally, this point is a general one. It transcends not only Jupiter but our solar system. When it comes to searching for life on exoplanets (Chapter 6), we generally think in terms of their potential habitability for Earth-type life. This is a pragmatic choice. While there may be life somewhere in the universe that's of an entirely different type, we can't search for an entity whose features are unknown. We could search for some *particular form* of non-Earth-type life, but in that case we'd have to specify its features first. Such an exercise would therefore start in the realm of speculation – something that's best avoided.

Although choosing to focus on the habitability of a planet for life that's broadly like Earthly life is primarily the result of practical considerations, it may be sensible from a more fundamental perspective too. There's a distinct possibility that most life across the universe is similar to that on our home planet – not in detail, of course, but in its general features. We'll explore this fascinating idea in Chapter 7.

The Habitable Zone

Earth is in the habitable zone of the Sun. This zone is defined by the possibility of there being liquid water on a planet's surface. The concept of the habitable zone is an old one. This name – which is the one that has stuck – was possibly first used by the Chinese-American astronomer Su-Shu Huang in 1959. Other names that have been used for the same concept include the Goldilocks zone, the temperate zone, and the liquid water belt. Implicit in all of these is a band, belt, or zone that, for a lone star, is of a fixed inner and outer margin, and therefore of a fixed width. It's thought of as circular, because in a typical planetary system such as our own the planets all orbit in the same plane – approximately, though not exactly – though in fact the habitable zone is potentially spherical.

The limits of a habitable zone are hard to specify exactly, but relatively easy to specify approximately. According to most sources, the Sun's habitable zone extends from a bit further out than the orbit of Venus to almost as far out as the orbit of Mars, so Earth is the only planet that's definitely within it. But because there's a little uncertainty about its limits, a case can be made for Venus and/or Mars to be hovering close to them. All the other planets of the solar system are unquestionably outside the habitable zone, even though Mercury is 'outside' it in the inner direction, while the others are outside it in both senses of that word.

Every star has a habitable zone, though in some cases this is complicated by stars being in pairs (binaries) or groups (multiple-star systems). In the simple case of lone stars like the Sun, the basis for the habitable zone is simple: the further away a planet's orbit is from its host star, the less radiation it receives and the colder it is. However, even without the complications of binary or multiple systems, there are other factors that need to be taken into consideration.

One of these other factors is orbital eccentricity – the degree to which a planet's orbit deviates from being a perfect circle. The orbit of the dwarf planet Pluto is very eccentric, with the result that while Pluto is further out from the Sun than Neptune most of the time, it's closer in for some of the time – in other words, the orbits cross over. There's no implication for the habitability of Pluto, because it's so far out from the Sun even when it's at its closest. But if Mars was inside Earth for part of its orbit, there would indeed be consequences for habitability. Although it's not, there may be equivalents of this phenomenon in other planetary systems.

Another factor that needs to be taken into account is a planet's atmosphere. The zone around the Sun where Venus orbits is colder than that within which Mercury is found. But that does not mean that the surface of Venus is colder than that of Mercury. In fact, at about 470 degrees Celsius, it's considerably warmer. This is because Venus has a thick atmosphere that impedes the loss of radiation, and thereby causes a runaway greenhouse effect. In contrast, Mercury hardly has an atmosphere at all – it's surrounded by gases whose collective density is so thin that on Earth we'd call it a vacuum. So there is no heat retention and no greenhouse effect. There is another type of effect that derives from Mercury's slow rate of rotation. A Mercurian day lasts for 59 Earth days, while a Mercurian year lasts for 88 Earth days. This means that most parts of the surface of this small planet experience long periods of unremitting sunlight separated by long periods of darkness, resulting in a pronounced difference in average temperature between what we might call the prevailing day-side and night-side. However, even the day-side is significantly cooler than Venus, which attests to the power of Venus's greenhouse effect.

Finally, we also need to take into account changes in time. Stars get hotter and brighter as they age. This means that habitable zones gradually move outwards. The reason for this is that as nuclear fusion proceeds in a star's core, more and more hydrogen is converted into helium. Of these two, helium is denser. As the core gets denser, it also gets hotter, and this speeds up the fusion process, with the result that more energy is produced per unit of time – a positive-feedback process. Although the rate of change is very slow, it has major effects over long enough periods of time. For example, the Sun will be hot enough in about two billion years to boil away all of Earth's oceans. That will mark the point at which the inner edge of the Sun's habitable zone has moved out beyond Earth's orbit.

So there are many complications, but despite these the habitable zone remains a key concept – many would say *the* key concept – in the search for life in the universe. It applies everywhere, from arm to arm within the Milky Way, and from galaxy to galaxy. It's the obvious way to focus our search for inhabited exoplanets – a focus that becomes increasingly necessary as we discover more and more exoplanets in general. It's a hugely exciting idea. But, as ever, we must temper excitement with caution. Mathematicians have a useful phrase: the 'necessary and sufficient conditions' for something to happen, or to be true. Let's apply this phrase to the habitable zone. First, we ask: is being in the habitable zone *necessary* for life to exist? We don't yet know, so the answer should be 'maybe'. Although surface water can't exist outside the zone, subsurface water can; we'll look at this issue in the next chapter. Second, we ask: is being in the habitable zone *sufficient* for life to exist? The answer this time is 'no'. This is because certain other criteria must also be satisfied, as we will now see.

Time for Evolution

Evolution is a slow process. To produce impressive results, it needs a considerable amount of time. As we saw in the previous chapter, evolution on Earth took about half a billion years to make microbes, more than three billion years to make large multicellular organisms, and about four billion years to make life forms whose intelligence had risen to the level where they were able to understand – and send and receive – radio signals. While rates of evolution may vary from planet to planet, there's no reason to believe they will do so by orders of magnitude. So it seems a reasonable guess that planets need to exist for billions of years rather than merely millions in order for evolution to get beyond unicellular life forms.

Let's imagine a planet that has satisfied all the criteria for a long lifespan that we've considered so far. First, it orbits a long-lived star. In practice, this excludes at least O and B class stars, and probably A class too. But this doesn't matter too much, as these three classes together make up less than 1% of all stars. Second, it's in the habitable zone. Third, it stays there, despite the possibility of orbital migration and the certainty of the habitable zone moving gradually outwards. These features *should* give a planet plenty of time for evolution to generate complex life forms, but they might not.

What's at issue here is the occurrence of drastic events that can stop evolution in its tracks. This has come close to happening on Earth several times, in what are called mass extinction events. The most recent and most famous of these (though not the most major) was the event in which the dinosaurs perished. This happened 66 million years ago, when an asteroid impacted Earth – specifically the Yucatan peninsula in what is present-day Mexico. It's thought that this impact and its climatic effects killed off not just the dinosaurs but about three-quarters of all forms of life.

However, there's a big difference between causing the extinction of most life and all of it. With most species gone but a sizeable minority remaining, evolution can in a sense fill the gaps. The proliferation of many kinds of large mammals in the wake of the dinosaurs' demise is a case in point. In a way, killing off many lineages stimulates evolution, while killing off all of them effectively presses a reset button, which requires another origin of life to get going again. Earth hasn't experienced a sufficiently large impact for this to happen, but it may do in the future. And many other planets might be impacted in this way at an earlier stage in their evolutionary process – for example, before the evolution of intelligence.

Drastic events can be initiated by the host star of a system as well as by wandering asteroids. The Sun is far from quiescent. Sudden surges of energy (flares) and matter (coronal mass ejections or CMEs) erupt periodically from its surface. Many of these are such that their volume exceeds that of the Earth. However, despite this, their effects on life here are small. There was a major geomagnetic storm in 1859 (called the Carrington event) caused by a flare and its associated CME. This caused widespread disruption to telegraphic communications. But short-term electrical disruption is a long way short of an extinction-level event. Some planets are less lucky in that they orbit stars that are prone to issuing forth flares and CMEs that make those of our Sun look tame. There are categories of stars called flare stars (quite common) and super-flare stars (quite rare).

For most planets orbiting most stars, extinction-level stellar eruptions and asteroid impacts are probably both infrequent events. But there's another reason why planet-wide extinction might occur that's much commoner. This is the phenomenon known as tidal locking, which is best explained from a starting point of the Moon.

We only ever see one side of the Moon, give or take a little bit of wobble. This could be called the visible side. The hemisphere (approximately) that can't be seen from Earth could be called the invisible side, but shouldn't be called the 'dark side' because all parts of the Moon get sunlight – the Moon is locked to the Earth, not to the Sun. To put it another way, the Moon exhibits synchronous rotation – its day and its year are effectively the same. If it were to spin more quickly on its axis, we would see all faces of the Moon. Equally, if it were to stop spinning on its axis altogether, we would again see all faces. If you find this hard to picture, draw a face on one side of a ball and hold it at arm's length while rotating your body; you can then experiment with the effects of different speeds of twisting the ball round on its own axis.

Ocean tides on Earth are largely due to the gravitational interaction between the Earth and the Moon. Likewise, the tidal locking of the Moon to the Earth is due to this interaction. The mechanics are complex, but here's a broad outline of what happens. Just as the Moon can pull the water of our global ocean towards one side of the Earth, so too the Earth can pull at the Moon in a way that distorts its shape a little, with slight elongation occurring in the direction of the Earth-to-Moon axis. This creates a sort of 'handle' that gravity then further acts on, with the effect that the Moon's speed of rotation on its axis slows down. In the long term, it ends up at the stable point – that of spin-orbit synchronicity, with the tidal bulge or 'handle' continuously pointing towards Earth.

This gradual locking process is a mathematical one that takes place given certain masses of the interacting bodies and a certain range of distances apart. There's nothing unique about the Moon in this respect. Nor is there anything unique about moons (plural); just as these can become locked to their host planet, planets can become locked to their host star. There are no cases of spin-orbit synchronicity for the planets of our solar system, but Mercury exhibits a related phenomenon called spin–orbit resonance – also gravitationally induced. Mercury spins on its axis three times for every two orbits of the Sun. As I said earlier, a Mercurian day lasts 59 Earth days, a Mercurian year 88 Earth days. Three times 59 is the same as two times 88, ignoring the rounding errors caused by my use of integers.

Complete tidal locking of planets, where a planet always presents the same side to its host star, *does* occur in lots of other systems. It is especially important in relation to habitability when the host star is a red dwarf (class M). This is because the mathematics of the situation are such that, for these low-mass stars, the habitable zone is typically entirely within the tidal locking zone, meaning that *any planets on which evolution may start to occur will end up slowing in their speed of rotation until one side permanently faces the star.* This will have major effects on evolution, and may well stop it in its tracks, producing a planet-wide extinction event. This is not inevitable, but it's very likely. We could perhaps imagine a situation in which the life on the dark side dies (note that 'dark side' *is* applicable here), as does that on the permanently irradiated side, but life around the join between the two somehow survives, leading to what might be called a bio-ring as opposed to a bio-sphere. However, models of the atmospheric dynamics of such a planet don't produce optimistic results for such a scenario.

What this means is that there is a question mark over the likelihood of life on planets orbiting red dwarfs. Unlike the problems for life evolving on planets orbiting very large stars, this problem for those orbiting very small ones has a large effect on any calculations we do to estimate the overall number of inhabited planets. This is because, while very large stars are rare, very small ones are common: about three-quarters of all stars are thought to be red dwarfs. So when we come to do such calculations (in the final section of this chapter) we'll have to do them twice – once assuming planets orbiting red dwarfs can be long-term homes for life, and once assuming they can't be.

From Habitable to Inhabited

So far in this chapter, we've focused on planetary habitability. But there's a big difference between being potentially habitable and being actually inhabited. What we know about the scale of the universe and the ubiquity of planetary systems puts the existence of large numbers of habitable planets beyond reasonable doubt. But that doesn't necessarily imply that there are equally large numbers of inhabited planets. I'd say it's strongly suggestive, but it's certainly not conclusive.

The question this takes us to is how likely life is to originate, given suitable conditions. To contemplate this issue, let's consider ten different young planetary systems, each centred on a single star that's of the same type as the Sun (class G). We'll imagine that each system has a single rocky planet in the habitable zone of the host star, and that these ten planets are all broadly similar to the ancient Earth, before life began here. In fact, we could have the solar system as one of our ten examples. In each case, we think about what will happen as time moves inexorably forward for millions and eventually billions of years from our young-system starting point.

On Earth, we know that life began after about half a billion years, give or take a wide margin of error, and that evolution has been going on without a pause ever since. We know that life began in an aquatic environment, but we don't know exactly which kind – warm pond, shallow sea, deep-ocean vent. We know that all today's species are descended from a common ancestor, but we don't know which one. And we know that life may have originated multiple times, but if so then all its origins but one eventually fizzled out, leaving no surviving descendants. We know that the successful origin of life began with haphazard collections of small organic molecules and ended up as organized and bounded collections of both large and small organic molecules – bounded collections (cells) that were able to reproduce themselves with at least a reasonable degree of fidelity.

The above paragraph is a brief summary of our discussion of the origin of life on Earth (Chapter 3), and is a reasonable account of the status of our current knowledge (and lack of it) on this subject. Does it help us to imagine what might happen on our other nine Earth-like planets, and if so, how?

There's a problem here, as follows. In later evolution on Earth, some innovations happened multiple times, thereby suggesting that the evolutionary transition concerned is in some sense 'easy'. For example, as we saw earlier, multicellularity has originated many times – between about 5 and 30, depending on how we define it. This suggests that if unicellular life evolved on other Earth-like planets, multicellular forms would eventually follow. But we can't use the same rationale for the origin of life itself, because of our uncertainty about the possibility of origins on Earth that fizzled out early and thus have no descendants in the present-day biota.

Another line of argument from what's happened here on Earth to what might happen elsewhere starts not from single versus multiple, but rather from early versus late. Even the most cautious scientists working on the origin of life on our home planet agree that organisms existed here by 3.5 billion years ago, in other words about one billion years after the Earth formed from the Sun's proto-planetary disc. One billion years into a planetary lifespan of about 10 billion is 'early' in proportional terms, even if not in absolute ones. Perhaps 'early', like 'multiple', can be interpreted as 'easy'? Perhaps on most or all Earth-like planets life originates sometime within the first two or three billion years, but with plenty of variation in the exact timing from one planet to another? The answer is embedded in the question: perhaps.

Considering where life originated on Earth rather than when is another way into the question of what happens on the other nine hypothetical Earth-like planets. But again our uncertainty precludes clear conclusions. If all that life needs in order to begin is a plethora of simple organic molecules in a body of water, such a concoction should be available on all our other Earth-like planets, since they're all in the habitable zones of their host stars. But if hydrothermal vents are necessary, then those planets might not be similar *enough* to Earth to have these habitats. For example, if they lack plate tectonics they will probably lack vents, since vents are typically found at mid-ocean ridges, which are places where new plate material is produced.

You can see the problem here. It's easy to imagine two extreme situations. In one, life originates on all ten planets at least within the first couple of billion years, if not earlier. In the other, life on Earth involved some unique combination of things that was vanishingly improbable and thus never happened on any of the other nine. In the first situation, ten habitable planets become ten inhabited ones. In the second, ten habitable planets become a single inhabited one plus nine that remain uninhabited. We don't have the ability to distinguish which of these is more likely, or indeed to decide that perhaps the truth lies somewhere in between. To my mind, this is the biggest unknown in the search for life in the universe. I suspect that the truth is much closer to the 'inevitable life' end of the spectrum of possibilities than to the alternative end – that of 'vanishingly improbable life'. Some scientists agree with me on this point, while others disagree. So the only sensible strategy when we try to calculate the number of inhabited planets in the universe is to use two very different probabilities for the origin of life and look at the effects these have on the guestimated numbers.

A Galactic Habitable Zone?

The habitable zone that's the focus of attention in planetary science, and the one we've been concentrating on so far, is the *circumstellar* one; in other words, it's a zone around a single star, or, in some cases, around a binary or multiple star system. However, some authors have proposed a parallel concept at a higher level – a *galactic* habitable zone. This is often pictured as having a broadly similar shape to its smaller circumstellar counterpart in that it's roughly circular. Sometimes it's described as donut-shaped. It excludes both a spiral galaxy's central bulge and its periphery.

This idea of a galactic habitable zone has been much criticized, and I find myself on the side of the critics. To me, it's an attempt to turn something minor into something major. There almost certainly is some variation in the probability of finding inhabited planets from one part of the Milky Way to another, and this is almost certainly true of other galaxies also, but there is no basis for turning this variation into the black-and-white binary switch of a 'yes' or a 'no' to life. Recall that the basis for the circumstellar habitable zone is the issue of whether water could exist on a planet's surface. There is no equivalent physical criterion that can be used to delineate a galactic habitable zone, and we shouldn't pretend that there is.

Naturally, those in favour of acknowledging the existence of a habitable zone at the level of a whole galaxy have some physical basis for their view – it isn't a complete flight of fancy. The central bulge is excluded from the zone because the density of stars there is so great. The idea is that with such dense packing the risk of disruption of planetary systems due to overly close supernova explosions is very high. And there's a comparable risk in some parts of the periphery too, despite the fact that the overall density of stars there is quite low. This risk arises because many of the stars in the periphery are found in globular clusters. These are large, tightly packed groups of typically very old stars, ranging in size from a few thousand to about a million per cluster. Not only do such clusters share the problem of the galaxy's central bulge in having dense star packing, but the fact that they're old stars means that their content of elements heavier than helium (their 'metallicity') is low, which causes problems for the origins of both rocky planets and life, as we saw earlier.

But variation in the probability of planets becoming inhabited between one large span of the galaxy and another, due to agencies such as the density of star packing, is just that. It's unhelpful to try to turn this into the basis of a claimed galactic habitable zone around which a boundary can be drawn. Moreover, stars and their accompanying planets move around in galaxies – they don't occupy fixed positions. Our solar system completes one galactic orbit about every 250 million years. When the first ever dinosaur appeared, Earth was roughly where it is now. During the heyday of these great reptiles, we were at the other side of the galaxy, and by the time they went extinct we were three-quarters of the way back to our current location.

In conclusion, the concept of a circumstellar habitable zone is useful, perhaps even essential, but the idea of a galactic equivalent is misleading and is best avoided. All parts of the galaxy are potential homes for planets with life. This is important for any attempt to scale up what we know of our local neighbourhood – the Orion spur – to the galaxy as a whole, which will be our quest in the following section. Recall that at the end of Chapter 2 I said that there were two perspectives on our 'local' search for life: as an end in itself, and as a sample of any arm of any spiral galaxy. But as a sampling exercise it's also a sample of the Milky Way. The lower probability of life in the central bulge of a spiral galaxy may mean that the Orion arm is a better pointer to comparable locations in other such galaxies (Andromeda, for example) than it is to our own galaxy's core. However, that's again an issue of probability, not a binary switch.

Millions of Inhabited Planets

Let's now build on what we've considered so far in this chapter. Here's a recap before we proceed. Taking a pragmatic approach, we search for Earth-like life, while keeping an open mind about the possibility of other types existing too. We focus on planets orbiting stars, while keeping an open mind about the possibility of life on moons orbiting those planets and life on rogue planets that lurk in the depths of interstellar space. We focus on surface life, while acknowledging that subterranean life exists on Earth, and might exist, even in the absence of any surface signs of life, elsewhere. We recognize that surface life can only exist in the habitable zone, where liquid water can be found on a planet's surface. We acknowledge that some planets in the habitable zone aren't good places for life for several reasons – for example because they are tidally locked to their host star. And we understand that not all habitable planets become inhabited ones.

We'll now proceed towards some guestimates of the commonness of life in the galaxy, and in the observable universe. The calculations depend on habitable planets existing right across the Milky Way and across most other galaxies, or at least most spiral ones. The basis of the calculations will be simple, starting with the number of orbiting planets of *all* types, and ending with the number of inhabited ones.

Our starting point, then, is the number of orbiting planets in the Milky Way. To the nearest order of magnitude, this is probably a trillion. There are thought to be between 200 and 400 billion stars in our home galaxy. Given that stars form with associated proto-planetary discs of gas and dust, from which planets form, it seems that just about all stars have planetary systems. We don't know the average number of planets per system, but it's probably in single figures. If there are 200 billion stars with an average of five planets per star, we get an estimate of a trillion planets in the galaxy. If instead there are 400 billion stars with an average of 2.5 orbiting planets, we also get an estimate of a trillion. Naturally, there are wide errors on this estimate, but perhaps not so wide as to change it by an order of magnitude.

Now we begin to narrow down via considerations of habitability. At each step, I'll suggest two figures, one more optimistic, the other more pessimistic. The first step is to come up with a figure for the number of planets in the habitable zone. In the solar system, this is just one – Earth. In some systems it will probably be zero. In other cases it will be 'several'; we already know of one case where it is three (TRAPPIST-1, Chapter 6). In fact, a fraction is more useful to our calculations than a number. So we have the solar system's 0.125 (one-eighth) and TRAPPIST-1's 0.429 (three planets out of seven). A reasonable guestimate of the overall fraction is 0.1 (10%), but I'll also use a more pessimistic 0.01 (1%). These fractions, and all the others discussed below, can be found in Figure 4.1, along with the end result of our calculation.

Starting point: About a trillion planets (10^{12})

	Optimistic	Pessimistic
% that are in the habitable zone	10	1
% of those that are habitable	5	1
% of those that become inhabited	90	10
Combined %	0.45	0.001
Number of inhabited planets in the Milky Way	4.5 billion (4.5×10^9)	10 million (10^7)

Figure 4.1 Steps in guestimating the number of inhabited planets in the Milky Way, from a starting point of there being roughly a trillion orbiting planets overall. At each step I use two figures – one optimistic, the other pessimistic. Using the optimistic estimates, there are about 4.5 billion inhabited planets in our galaxy. Using their pessimistic counterparts, there are 'only' about 10 million. The truth is probably in between these two figures. Even if it's closer to the lower one, there are a lot of inhabited planets in the Milky Way – and probably in most other galaxies too.

Next, we need to come up with a figure for the fraction of planets within habitable zones that are habitable. For this, we need to exclude all planets orbiting short-lived stars, all that are too young for life to have evolved or too old for it to remain in the face of increasing luminosity of their host stars, some or all that are tidally locked, and perhaps all that orbit stars prone to massive flares. There are too many variables here to be confident of the overall fraction of planets to exclude, and thus conversely the fraction on which to focus. Because of this, even the more optimistic of the two figures I'm going to use is low: 0.05. The more pessimistic figure is 0.01. This involves effectively excluding all planets orbiting red dwarfs, because of the common problem of tidal locking there.

Now we conclude with the most tricky question of all: what fraction of habitable planets become inhabited? Recall that there's a spectrum of views on the likelihood of origins of life on potentially habitable bodies. At one end there's the 'inevitable life' perspective; at the other there's its 'vanishingly improbable life' counterpoint. I'm going to represent these with guestimates of 0.9 and 0.1 respectively. Clearly, which one we choose will make an almost 10-fold difference to the resulting number of anticipated planets with life.

Finally, we're in a position to do the necessary calculations. Using all the more optimistic figures, we get a guestimate of 4.5 billion planets with life in the Milky Way. Using the more pessimistic ones, that is reduced to 10 million. The truth is probably somewhere in between the two. To scale these numbers up from the galaxy to the observable universe, we need to multiply by the number of galaxies, or, for a more pessimistic scaling, just the number of spiral ones. The appropriate multiples are, respectively, a trillion and a quarter of a trillion – since spirals are thought to represent about a quarter of the total. Applying the more optimistic scaling to the higher of our two guestimates for the number of Milky Way planets with life, we get 1 trillion times 4.5 billion, which is 4.5×10^{21} (a sextillion). Applying the more pessimistic scaling to our lower guestimate, we get 2.5×10^{18} (a quintillion). Either way, it's a mind-boggling number of inhabited planets.

Before moving on to other topics, it's worth stressing that all the above calculations are for life in general, not intelligent life in particular. There are probably lots of planets with nothing more complex than microbes – these may even be the majority. But some planets will have multicellular life forms, and some of those will include intelligent life. We'll focus on intelligent life in Chapter 8.

5 Life in the Solar System

A Unique Opportunity

The thought of there being millions of planets with life in the observable universe is inspirational. But it's only that – a thought. Or perhaps a bit more than that – a thought with probability on its side. But the gap between probability and certainty is a huge one. We won't really *feel* the presence of extraterrestrial life until we know for sure that it's there. So we need evidence. I started the book with a look at a paper that focused on the need for a cool assessment of evidence and the importance of not jumping to conclusions. In this chapter we'll return to that issue.

Acquiring evidence logically precedes assessing it, and the solar system provides a unique opportunity in this respect. It's the only system that our current technology allows us to inspect by flying space probes to its various planets and moons to get a close look. How close depends on the probe. Some simply fly close to a planet or moon on a single occasion – flybys. Others spend longer, making multiple trips around the body concerned – orbiters. Some make descents to the surface. These include both crash landings – impactors – and controlled ones – landers. Of the latter, some are capable of powered movement – rovers. None of these options yet exists for any other planetary systems, though there's an ambitious plan to laser-propel tiny 'nano-craft' to the nearest system to our own – Alpha Centauri (Chapter 6).

Our progress in flying spacecraft to various parts of the solar system has been amazing. The first ever spacecraft was the Soviet *Sputnik 1*, which orbited Earth more than 1000 times from October 1957 to January 1958. The first lunar impactor was *Luna 2*, which crashed into the Moon in September 1959. The first spacecraft to reach the surface of another planet (Venus) were *Venera 3* (impact, 1966) and *Venera 7* (soft landing, 1970). From these early starts, we've progressed to having multiple rovers on Mars, landing the *Huygens* probe on Saturn's largest moon Titan, and landing probes on both asteroids and comets. NASA's *New Horizons* craft did a flyby of Pluto in 2015, revealing a curious heart-shaped pattern on its surface. And the *Voyager* twins (*1* and *2*), launched in 1977, have now reached the edge of the solar system.

However, along with the unique opportunity for physical exploration, the solar system also provides us with a unique challenge. Earth is the only planet that's squarely within the Sun's habitable zone – though Venus and Mars come close. And our sterile Moon is the only natural satellite in the zone. So, to look for alien life in the solar system, we're forced to look either beyond the zone or back in time. This takes us to the possibilities of present-day life in strange places – such as Venusian clouds or Europan subsurface seas – and the distant past, when water flowed on the surface of Mars. Focusing on the solar system probably also restricts us to the possibility of microbial rather than animal or plant life, but a useful caveat about the assumption of 'microbes-at-best' can be found at the end of the fascinating 2013 sci-fi film *Europa Report*, directed by Sebastian Cordero (no spoilers!).

Distances from the Sun

The solar system contains eight planets, several dwarf planets, and more than 170 moons, plus countless asteroids and comets. Almost everyone knows the order of the planets, starting with Mercury and ending with Neptune. And many people also know where most of the asteroids and comets fall in this order – between Mars and Jupiter, and beyond Neptune, respectively. But there's a big difference between an ordinal scale, in other words a simple sequence, and an interval scale, in other words a series of measurements of the actual distances (or 'intervals') separating the various objects in the sequence. This is especially true of the solar system, where most simple diagrams show the planets artificially close together and approximately equally spaced.

We need to get beyond such a simplified picture to understand the true scale of the system; and to do this we need a unit that works at the scale concerned.

Miles or kilometres are too small. And light years are too big, because the whole system fits well within a radius of a single light year (with the proviso that our definition of the solar system excludes the remote Oort cloud). The intermediate-length unit that's used instead is the astronomical unit (AU). One AU is the distance from the Earth to the Sun, that is, about 150 million kilometres or 93 million miles. Figure 5.1 shows the distances to scale, and the approximate number of AU each planet is from the Sun, though bear in mind that these numbers are variable, because the orbits are elliptical rather than circular. The approximate positions of the edges of the habitable zone are also shown in the figure. Notice how far away from the zone is the Saturnian system, where at least one moon – Enceladus – is thought to be a possible home to life.

To date, there have been no credible claims of life on the innermost planet, Mercury, and the same is true of the two outermost ones, Uranus and Neptune.

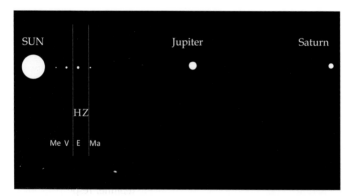

Figure 5.1 Positions of the inner six planets of the solar system, shown approximately to scale. The distances out of the orbits, from Mercury (Me) to Saturn, are about 0.4, 0.7, 1, 1.5, 5, and 10 astronomical units (AU), with 1 AU being defined as the distance of the Earth from the Sun. The two outermost planets are not shown. The orbit of Uranus is about 20 AU from the Sun, that of Neptune about 30 AU – so that at the scale used in the illustration, Neptune would be about two page-widths off the right-hand margin. Note the position of the habitable zone (HZ), with Earth (E) squarely within it. The edges of the zone are hard to specify exactly, but are probably such as to exclude Venus (V) and Mars (Ma).

Asteroids are unlikely places for life due to their lack of both atmospheres and liquid water. The same is true of comets, though the situation is more complex with these 'dirty snowballs' because they have plenty of H_2O. However, this exists in solid form – ice – within the comet's nucleus. Some of this sublimates to water vapour as a comet approaches the Sun, meaning that it goes straight from solid to gas, without passing through the liquid phase that's normally in between the other two.

In the next few sections, we'll look at possibilities of life on various solar system bodies, excluding what we might call the 'hopeless cases', namely those listed in the previous paragraph. We'll take the 'hopeful' ones in order from proximal (closest in) to distal (furthest out). So we'll start with Venus, orbiting at a mere 0.7 AU from the Sun, and finish with the Saturnian system, orbiting more than ten times further out, at around 10 AU. Where applicable, we'll look at any claimed evidence for life, and will assess where it falls on the spectrum from tenuous to conclusive.

The Atmosphere of Venus

Venus is often described as having a hellish environment. As we've seen, its average surface temperature is even higher than that of Mercury. Its surface pressure is about 100 times that of Earth. And there's sulphuric acid rain. So the idea of life on the Venusian surface seems like a non-starter. But a paper in *Nature Astronomy*, first published online in 2020, raised the possibility that there might be microbial life forms lurking in Venus's atmosphere, where conditions are less forbidding. The authors – British astronomer Jane Greaves and her colleagues – reported the apparent discovery of the gas called phosphine in Venusian clouds. Recent studies have refuted this claim, and it now seems likely that there's no phosphine on Venus at all. However, the possibility of there being phosphine on Venus provides a useful opportunity to look at how we assess evidence for alien life.

Phosphine is not an organic compound. It's a mixture of phosphorus and hydrogen (PH_3). It's highly toxic. It's found in other planetary atmospheres, including that of Jupiter. So why was its apparent discovery on Venus exciting from the perspective of alien life?

Chemicals in planetary atmospheres can be produced by biotic or abiotic means. In other words, they can be produced by living organisms, or by chemical processes that don't involve life forms of any kind. Jupiter's phosphine is abiotic in origin. Its existence there hasn't caused a stir among astrobiologists. But the nature of Venus's atmosphere (dominated by carbon dioxide) is sufficiently different to that of Jupiter (dominated by hydrogen) that phosphine should not be produced there by purely chemical means. This, coupled with the fact that some microbes on Earth produce phosphine, were the bases for the excitement about the claimed Venusian find.

The authors of the paper in *Nature Astronomy* were very careful not to conclude that the phosphine on Venus was biogenic. They pointed out that this was just one of several possibilities. But when the paper was reported in the popular media, all the other possibilities sank from sight, and the issue of possible Venusian life came to the fore. This is exactly the sort of distortion of scientific evidence that we need to avoid. So here we return to the scale of conclusiveness of scientific evidence for alien life that I mentioned at the start of the book, namely the one proposed by James Green *et al.* in 2021. The authors called it the CoLD scale, an acronym of Confidence of Life Detection. They gave it seven points, from 1 (low confidence) to 7 (high). However, they made it clear that their proposal was intended to initiate discussion on the issue concerned, not to provide a final answer. And doubtless others who follow their lead will use a greater or smaller number of points.

Venusian phosphine is right at the bottom of the scale, whatever exact form it takes. This is the point at which something has been detected that *could* result from biological activity. To get to higher levels, the thing (or signal) detected needs to be shown to be real rather than due to contamination or misinterpretation; also, non-biological explanations have to be progressively ruled out. Neither of these criteria is met in relation to phosphine on Venus.

Many readers of the paper on Venusian phosphine probably came to the same conclusion that I did: that there was some abiotic way of producing tiny quantities of phosphine, and that no living organisms were involved. After all, to explain the existence of even a single species of microbe in the clouds of Venus, it would be necessary to accept either: (a) that an evolutionary process had taken place there over millions of years; or (b) that space-wandering microbes had somehow ended up in the Venusian atmosphere, a case of so-called 'panspermia', which we'll look at in general terms in Chapter 7. Both of these seem extremely unlikely.

However, the truth was in a sense even more negative: the apparent detection of phosphine probably wasn't real. The NASA astronomer Geronimo Villanueva and his colleagues subsequently showed that there is no phosphine in the atmosphere of Venus (or at best a vanishingly small amount of it, less than one part per billion), and that the signal interpreted as phosphine may have been sulphur dioxide. Given that there's no phosphine, there's no evidence for life. Thus, we return to our original view of Venus as being an entirely sterile planet – not just its surface, but its atmosphere also.

Mars in Deep Time

Present-day Mars may be sterile too. Attempts to find evidence of extant life there have so far failed – and there have been quite a few of them. Admittedly, there were some early positive signs, notably the 'biology experiments' conducted by the *Viking 1* and *2* landers, which arrived at the red planet in 1976. But these were flawed, as became obvious when the apparently positive result for metabolizing life in the Martian dust was also obtained from a control treatment using sterilized material. We're about as certain as we can be that there's no life today either on the surface of Mars or in its atmosphere. The presence of subterranean life is harder to exclude, but there's no evidence for it as yet.

However, Mars in the distant past was a very different place to its current counterpart. We have extensive evidence for water flowing across the Martian surface in ancient times. How ancient? It's hard to be sure of the period through which Mars had large bodies of surface water, but the consensus is that such water bodies existed from before 4 billion years ago to sometime approaching 3.5 billion years ago. This corresponds very roughly to the time when life is thought to have originated and begun its evolution here on Earth. Perhaps life on the two planets ran in parallel for this immense stretch of time, only for Martian evolution to stop in its tracks at a point where all it had produced were microbes?

There's a problem with this scenario. Since the Sun was significantly cooler back then than it is today, Mars should have been even less habitable then compared to now, other things being equal. But they weren't. It seems that early Mars had a thick atmosphere, as present-day Venus does, and that this caused a runaway greenhouse effect that heated the planet's surface and allowed the existence of liquid water. There are many valley systems on the Martian surface that seem to represent ancient river systems; and some craters, including Gale and Jezero (landing sites of the *Curiosity* and *Perseverance* rovers respectively), that may have been ancient lakes.

It's worth remembering that bodies of liquid on the surface of a planet or moon needn't be bodies of water. The large lakes on the surface of Saturn's moon Titan (next section) are a mixture of liquid hydrocarbons. However, the mineral haematite, typically formed in aqueous solution, is found on Mars, which points to water. Some of this mineral is found in the form of spherules, sometimes referred to as 'blueberries'. As an aside, one wacky 2021 paper interpreted these spherules as fungi. This, needless to say, is regarded by the rest of the scientific community with incredulity.

So, it seems that back in 'deep time' – an expression much used by evolutionary biologists – Mars had a thick atmosphere, was warmer than today, and had surface rivers and lakes, and possibly even an ocean in its relatively flat northern hemisphere. If life evolved there, we should be able to find some signs of it. These would not, of course, take the form of the kind of fossils familiar from Earth's evolutionary past, such as trilobites or dinosaurs, because if life existed on Mars it became extinct at far too early a stage for animals to have evolved. But there should be some kind of fossils in the broad sense, whether morphological ones representing lithified single cells, or biochemical ones.

The only way to look for such fossils is to seek them *in situ*, as the *Perseverance* rover is doing right now. We can't expect them to come to Earth of their own volition. And yet, strangely, just such a scenario was suggested by a group of scientists that examined a Martian meteorite found in Antarctica in 1984. This meteorite is given the code name ALH 84001, which incorporates three pieces of information: the place it was found at (the Allan Hills), the year it was found (1984), and the fact that it was the first meteorite found in that place in that year

(001). The meteorite is about 10 centimetres long, and weighs just under 2 kilograms; it's thought to be about four billion years old. The team studying it took a chip from the meteorite and broke it along pre-existing fracture lines, revealing fresh surfaces that weren't obscured by the fusion crust that a meteorite acquires as it burns its way through our atmosphere. One of these fresh surfaces had features that were interpreted by the team involved as being fossilized bacteria-like cells. At the time there was much scepticism; now the hypothesis has been almost universally rejected because the 'biogenic' features can be explained in purely physical and chemical terms.

This is another point at which we should consider the nature of proposed evidence for extraterrestrial life. Like the claims for life in the Venusian clouds, this purported evidence for past life on Mars is not persuasive. On the CoLD scale of evidence quality it scores higher than the phosphine, but not by much. Yet when the paper claiming that the features of ALH 84001 were fossilized Martian bacteria was published in 1996, the story was sensationalized in the popular media. President Bill Clinton even referred to it in one of his speeches.

It's worth rehearsing the series of events that must take place for a fossil-bearing Martian meteorite to arrive here on Earth, in order to understand just how improbable the whole thing really is. Let's suppose that life originated and began to evolve on Mars at about the same time as on Earth. Let's further suppose that bacteria-like organisms quickly spread across the planet shortly after their origin, as probably happened here on Earth. When colonies of bacteria died, they mostly rotted away to nothing, as happens here. But in a few cases their forms were preserved, as on Earth – the earliest stromatolites here *may* be fossilized cyanobacteria or other microbes.

Even if this is all true, we can ask what proportion of rocks from the era concerned would contain microbial fossils. On Earth it's tiny, and on a hypothetical ancient Mars that was probably the case too. So the chances of a small meteorite that arrives on Earth from Mars actually having constituent fossils is very low. Now imagine what they have to survive to get here. Meteorites from Mars are rare because they result from occasional instances of *other* meteorites arriving on Mars (probably from the asteroid belt) and causing an impact explosion, which sends bits of Mars itself up with sufficient force that they escape its atmosphere. Did ALH 84001 have fossils of Martian life? No. Was there Martian life four billion years ago? Maybe. We shouldn't let the hype associated with particular purported evidence for life cause us to reject not just the evidence but life itself. Let's wait and see what, if anything, *Perseverance* discovers.

Moons of Jupiter and Saturn

Up to now, Mars has been the prime focus for spacecraft whose mission involves a 'searching for life' component, but that may change in the relatively near future, in favour of certain moons of the gas giants Jupiter and Saturn. These planets and their moons were originally thought to be so far out from the Sun – orbiting at about 5 and 10 AU, compared to Earth's 1 AU and Mars's 1.5 – that a search for life would be fruitless. The planets themselves lack solid surfaces, while those of the moons are frozen solid.

But our view of these moons changed markedly as a result of the *Galileo* and *Cassini* missions. *Galileo* studied the Jovian system from 1995 to 2003, while *Cassini* studied the Saturnian one from 2004 to 2017. As a result of data collected by these spacecraft, we came to understand that some of the moons of Jupiter and Saturn have vast subsurface oceans of liquid water. But what sort of data? After all, neither of them had the capacity to drill through several kilometres of surface ice to inspect what's underneath. The *Cassini* mission did include a lander – the *Huygens* probe – which made a controlled descent to the surface of Saturn's largest moon Titan, landing in January 2005. However, the furthest that *Huygens* penetrated the surface was about 10 centimetres, as opposed to the 10 kilometres or more that might be required to sample a subsurface ocean.

Although Titan may well have such an ocean, the best evidence for large subsurface bodies of water so far comes from a smaller Saturnian moon – Enceladus – and from the Jovian moon Europa. Plumes of material can be seen spewing out from the south polar region of Enceladus. They emanate from cracks in the surface ice that are often referred to as 'tiger stripes'. *Cassini* was able to fly through these plumes and sample them. They turned out to be primarily composed of water-ice. What seems to be happening is that liquid water explodes out of weak points in the surface, turning to crystals of ice as it enters space, which it does as soon as it emerges, given that Enceladus has no atmosphere.

Europa seems to exhibit plumes too, but they're less prominent, and more intermittent, than those of Enceladus. This Jovian moon has an icy surface that is covered in cracks. A region of it that is multiply cracked in many directions is referred to as Conamara Chaos, after the western Irish region of Conamara (English spelling Connemara), with its complex patchwork of land and lakes. Various patches on the surface of Europa are described as ice rafts. These vary in size, but typically consist of a few square kilometres. The idea is that the surface gets broken up into slabs of ice that float on the underlying water before re-freezing.

The inference of subsurface oceans is not based on plumes and rafts alone. Much mathematical modelling work has been conducted, and it points not just to a few subsurface lakes – like Lake Vostok under Antarctica's ice – but rather to global oceans of considerable depth. And such oceans may not be confined to Enceladus and Europa. Other icy moons may have them too. Prime candidates are Jupiter's Ganymede – the largest moon in the solar system – and Saturn's Titan. There may even be subsurface oceans on some of the moons of Uranus and Neptune.

Now we need to confront the elephant in the room. How is liquid water possible so far from the Sun? The answer to this question lies in a simple phrase with a complex meaning: tidal heating. We've met 'tidal' (meaning gravitational) processes already – when we looked at tidal locking in the previous chapter. Now they appear again, but in a different guise. As a moon orbits its host planet, changing gravitational forces cause internal friction, which in turn causes an increase in temperature. This is sufficient to prevent water beneath the surface from freezing – hence the subsurface oceans.

Titan is particularly interesting from an astrobiological perspective, because it may have a subsurface ocean, but it also has surface lakes. As I mentioned in the previous section, the liquid in these isn't water. Rather, it's a mixture of hydrocarbons. The main constituents are ethane and methane, with smaller amounts of various other chemicals, including propane and hydrogen cyanide. Thus, there are two possible kinds of life on Titan. One would be Earth-like life that's exclusively deep down below the surface. The other might be some completely alien biochemistry, but in a superficially familiar-looking surface lake. The largest of Titan's lakes are so large that they're usually called seas. The most extensive of all is the Kraken Mare, which is larger than Asia's Caspian Sea. Which of these two types of life is more probable? I'd guess that it's the subsurface aqueous type, because the extremely low temperatures of the surface lakes would present a big obstacle to 'metabolism' of any kind.

There are exciting space missions planned to the Jovian system, namely ESA's *JUICE*, a clumsy acronym of Jupiter Icy Moons Explorer (*JIME* would have worked better, I think), due to launch in 2023, and NASA's *Europa Clipper*, scheduled to launch in 2024. There is also a plan for a privately funded mission to the Saturnian system – to Enceladus in particular. This is part of the Breakthrough Initiatives, launched in 2015 by the Russian entrepreneur–physicist and billionaire Yuri Milner, with support from many leading scientists, including Stephen Hawking. But we'll have to wait a while for results. Even if the first of these missions to launch is on schedule, it won't arrive at its destination until 2031.

Visitors to Our System: 'Oumuamua

I started this chapter by emphasizing the fact that we can study the various bodies of the solar system 'up close', something that's not possible for other planetary systems. However, occasionally bodies from other systems pay us a visit. There have probably been many such interstellar visitors over the course of our planetary history, but most have gone unobserved. However, in humanity's space age, observation is possible. Two such objects have been observed in the last few years – one in 2017 and one in 2019. The later arrival was a comet (called Borisov, after the Crimean amateur astronomer Gennadiy Borisov who discovered it). This comet was ordinary in most ways except for its out-of-system origin. But the earlier visitor could instead be described as extraordinary.

The 2017 visitor was named 'Oumuamua, which is a Hawaiian word that roughly translates as 'first messenger from afar'. It was discovered by Canadian astronomer Robert Weryk, using a telescope on the Hawaiian island of Maui. It arrived in the solar system 'from above', in the sense of coming from Earth's north. Its direction of entry suggests that it came from the area of space defined by the northern constellation of Lyra. It passed through our system on what's called a hyperbolic trajectory, meaning that it won't be captured by the Sun's gravity and end up as part of our system; rather, now that it has passed its closest point to the Sun, it is heading back into interstellar space, though in a different direction from its original one. The fact that its course was changed by its encounter with our Sun means that its 'original trajectory' may not have been that at all. It might have crossed through other planetary systems before ours, and been redirected by their suns just as it was by ours.

But what exactly was 'Oumuamua, and why am I spending time discussing it here, when I dismissed comet Borisov in a couple of lines? Well, to begin with, 'Oumuamua didn't have a tail, and wasn't a comet. It's sometimes described as an interstellar asteroid, but that's not a universally agreed status. It's very unlike a typical asteroid in its shape. It seems to be long and thin, perhaps a few hundreds of metres long and a few tens of metres wide, whereas most asteroids are irregular and squat. It is referred to as an interstellar *object* rather than an interstellar asteroid, which leaves open its classification.

Figure 5.2 shows two features of 'Oumuamua, one observed, the other based on artistic licence. At the top is the observed 'light curve', that is, the amount of light reflected by the object over time. This is consistent with an elongate object (bottom) with a tumbling motion, though other interpretations cannot be ruled out. Most scientists think that 'Oumuamua is a rocky object, but Harvard astronomer Avi Loeb has argued the case for it being a material manufactured by an alien technology in his book *Extraterrestrial: The First Sign of Intelligent Life Beyond Earth*. There are several bases of Loeb's argument, perhaps the most persuasive being the fact that 'Oumuamua accelerated away from the Sun in a way that's hard to explain in terms of natural processes. Personally, I'm keeping an open mind. Unfortunately, 'Oumuamua is now getting further away all the time, and is thus harder to study. Its closest approach to the Sun – which was inside the orbit of Mercury – happened in September 2017. It's now in the outer reaches of the solar system. But there will surely be further interstellar visitors, and the debate over 'Oumuamua's identity suggests that we should study them intensively when they arrive. And evidence for past visitors from afar may yet emerge from reanalysis of data already collected. Recent analyses suggest that a small meteorite that reached Earth in 2014 was probably of interstellar origin.

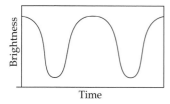

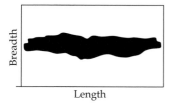

Figure 5.2   Observation (top) and speculation (bottom) on the 2017 interstellar visitor to our solar system called 'Oumuamua. The upper picture is the light curve of this object, showing a cyclic pattern of brightening and darkening. This is consistent with a long thin object that has a tumbling motion. The brightness varies considerably – by about a factor of ten. The time between one trough to the next suggests that 'Oumuamua rotates on its axis about every four days, though there is considerable uncertainty about this estimate. The lower picture is an artist's impression of what the object might look like if it's made of rock. The other interpretation is that the object is an alien artefact – perhaps a spacecraft, or a discarded piece of one.

Conclusions: How Many Inhabited Bodies?

We've now looked at the possibilities of life on Venus, Mars, and some of the icy moons of Jupiter and Saturn. There have been various suggestions of life, but none of them generally accepted. My own view is that there's no present-day life on Mars, or in the clouds of Venus. I remain open-minded about the possibility of life in subsurface oceans, for example those of Europa and Enceladus – and about whether 'Oumuamua was an object made by alien technology.

Where does this leave us in relation to the number of possibly inhabited bodies in the solar system? Let's put 'Oumuamua to one side, as it's *en route* to exiting our system altogether, and will probably never return. Also, let's put asteroids and comets to one side, since despite their large numbers it's probable that none are inhabited due to the lack of both liquid water and atmospheres. So our question relates to the 'large permanent residents' of the system, namely planets, dwarf planets, and moons. Even the majority of moons must be discarded as possibilities, given that their diminutive size renders them rather like asteroids. I'd include in this sack of discarded moons both those of Mars – Deimos and Phobos – and many of the small moons of the gas and ice giants.

This discarding leaves us with the following list of bodies. First of all, the familiar eight planets. Some scientists argue that there's evidence of a large 'planet 9' lurking in the outer reaches of the solar system, but most are sceptical of this hypothesis. The number of dwarf planets remains to be agreed, but is around ten, give or take a few. They include bodies in the asteroid belt (Ceres), bodies in the Kuiper belt (e.g. Pluto) and bodies whose orbits take them even further out than the Kuiper belt (e.g. Eris). The number of moons that are large enough to be spherical is as follows: one (Earth's) for the four rocky planets, four for Jupiter (the Galilean moons, which are far larger than all the others), seven for Saturn (including Titan and Enceladus), five for Uranus, and one for Neptune (Triton). Adding these up gives 18 'large round moons'. Adding this to the combined number of planets and dwarf planets, which is coincidentally also 18, gives a figure of 36. No doubt this figure might wobble a bit with future discoveries, but probably not by much.

The number of inhabited bodies in the solar system is thus somewhere between one and about 36. Realistically, I'd say the possible range is more like from one to ten, since many bodies can be reasonably excluded, for example, Mercury, Jupiter's ultra-volcanic moon Io, and everything in the Kuiper belt and beyond. We might even draw the line at about five – Earth plus the two most promising moons of the Jovian and Saturnian systems (Europa and Ganymede, Enceladus, and Titan). Even so, there's a big difference between one and five inhabited bodies. We won't know the right answer for at least a few years, and probably longer. My personal guess is 'one'. I suspect that Earth is the only inhabited body in the solar system. I'm a pessimist in this respect. But in contrast, I'm hugely optimistic about finding life elsewhere in our galaxy, which is where we're going next.

6 Life in Other Planetary Systems

The Excitement of Exoplanets

I introduced exoplanets – planets beyond our solar system – in Chapter 1, discussed them in the context of planetary systems in Chapter 2, and considered them in relation to the concept of habitability from a general perspective in Chapter 4. But so far I've given little detail about them. How many exoplanets, or exoplanetary systems, have I mentioned so far? Very few. That's about to change, but not in the sense of replacing a dearth of detail with a wealth of it. Rather, I'll be very selective about the particular exoplanets I discuss. This is essential, given that the number now known is huge, and most of them are irrelevant to the search for alien life.

Let's start with a timeline of exoplanet discovery, focusing on those findings that may be important with regard to looking for life. The timeline is short, because at the start of the year 1990 no exoplanets were known. Now, a mere thirty-something years later, the number has risen to over 5000. But it's necessary to take 'known' with a pinch of salt. When there's initial evidence that's suggestive of a planet orbiting a particular star, it's described as a 'candidate'. Later, when there's enough evidence that it can be regarded as conclusive, the exoplanet is described as 'confirmed'. In general, I'll stick to confirmed ones, though progress from candidate to confirmed may be gradual rather than sudden.

The location of the first exoplanets discovered wasn't at all conducive to their being habitable – quite the opposite, in fact. Their discovery was announced in a 1992 paper by the astronomers Aleksander Wolszczan and Dail Frail, who were based at the Arecibo observatory in Puerto Rico. What they found were two planets orbiting not a 'normal' star, but a stellar remnant (or 'dead star') called a pulsar. This is a rapidly rotating neutron star – a star whose atomic structure has collapsed – that's named after the intense beams of radiation it emits as it rotates, which from our perspective on Earth come across as a series of pulses, in the same way that the light from a lighthouse does. This was a hugely exciting discovery for astronomy, but not so much for astrobiology – astrobiological excitement came later, as we'll see shortly.

The next milestone was the discovery of an exoplanet orbiting a 'living' star rather than a 'dead' one. This was reported in a 1995 paper by the Swiss astronomers Michel Mayor and Didier Queloz, who were awarded the Nobel Prize for physics in 2019. The planet concerned is 51 Pegasi b. It's at this point that I need to explain a bit about the naming of exoplanets. If you already know about this, please skip to the next paragraph. There are several naming systems in place. I'll only explain those that we need, and I'll do so when we first encounter each of them. Here, we're encountering what I call the 'star-constellation-planet' system, another example of which is 55 Cancri e. Pegasi means 'in the constellation of Pegasus' (the winged horse), while Cancri means 'in the constellation of Cancer' (the crab). The numbers – 51 and 55 in these examples – identify a particular star in the constellation concerned. The letters – here b and e – specify a particular planet orbiting that star. The star is considered to be 'object a' in the system, and the planets are given letters from b onwards in an order that reflects their orbital distance from the star and/or their order of discovery.

Like the earlier discovery of the first pulsar planets, the discovery of 51 Pegasi b was hugely exciting from the perspective of astronomy and planetary science, but again it isn't a place that looks to have any chance of being habitable. It's a type of planet now known to be common, but then rather unexpected – a 'hot Jupiter'. This name is given to very large planets (Jupiter mass and above) orbiting very close to their host star. The orbit of 51 Pegasi b is at about 0.05 AU; this is roughly one-tenth of Mercury's distance from the Sun. Hot Jupiters are *very* hot. They cannot be homes to life as we know it, and I very much doubt that they are homes to any kind of life at all.

After nearly two decades of discovery of hot Jupiters and various other types of uninhabitable planets, the first Earth-sized planet discovered in a star's habitable zone was Kepler-186 f. This was reported in a 2014 paper in *Science* by the American astronomer Elisa Quintana and her colleagues. I'd rate this as the first truly exciting exoplanet discovery from an astrobiological perspective. But again we need to make a brief digression to understand the name of the planet, because this is a different naming system from the constellation-based one. This time, the system is 'telescope-star-planet'. So this is planet f (same rationale as before) orbiting 'star 186', as observed by the Kepler Space Telescope. There's no need to give a constellation here, because this telescope had a very restricted survey area, centred on the constellation of Cygnus the swan, and in particular one of its 'wings', though also overlapping parts of two other constellations, namely Cassiopeia and Draco.

Discoveries of further habitable-zone planets followed, and of course continues apace. There's no need to list them all. For now, I just want to give one more milestone of exoplanet discovery that's important to the search for life. This is the finding – mentioned briefly in Chapter 4 – of a system that appears to have not just a single planet in the habitable zone, but three of them: the TRAPPIST-1 system. In total, this has seven confirmed planets (labelled b to h), of which at least e, f, and g appear to be in the habitable zone. The discovery of this system was announced in papers published in 2016 and 2017 by the Belgian astronomer Michaël Gillon and his colleagues. They were using a telescope called TRAPPIST, or, to give it its full name, the Transiting Planets and Planetesimals Small Telescope, which is located at the La Silla observatory in Chile. The '1' refers to the first star it discovered with exoplanets orbiting it, so this naming system is essentially the same as that used for discoveries by Kepler.

This is an appropriate point to delve further into the role of transits in exoplanet discovery. TRAPPIST isn't the only telescope in whose name 'transit' is embedded. We saw another case in Chapter 1 – TESS, or Transiting Exoplanets Survey Satellite. There are several ways of detecting exoplanets, of which the transit method has been the most successful. Here's a very brief account of how it works. A telescope is trained on a particular star. Its instruments measure the amount of light that's incoming from that star over a protracted period – say a year. If a planet transits across the 'front' of the star (from our perspective), it blocks a small fraction of the light that's coming in our direction. If a dip in the amount of light is found on a regular basis – say every three months – then we conclude that it's caused by a planet with an orbital period (or 'year') that lasts for about a quarter of an Earth year.

This method is brilliant in its conceptual simplicity. However, it has certain limitations. In particular, it can only detect planets that belong to systems that are edge-on (or nearly so) from our vantage point, which is a small minority. Also, in common with other methods of detection, it finds big planets more easily than little ones. The smaller the proportion of light blocked by the planet, the less likely it is to be found using this method. So even when a system is edge-on, use of the transit method may find only its larger planets. This means that the number of planets currently listed for any particular exoplanetary system should be regarded as a minimum rather than a definitive final count.

This brief description of the transit method that has been responsible for the detection of so many exoplanets helps to stress the fact that most confirmed exoplanets have never actually been seen. Seeing the planet itself, and being able to photograph it, has been a rare thing so far. However, this will change with time, as 'direct-imaging' space telescopes are developed. In Chapter 1, we noted one of these that's at an advanced stage of planning – LUVOIR. However, that won't launch until the 2030s, so there's a long time to wait. For me, part of the excitement of exoplanet discovery to date lies in the fact that the methods used allow us to be certain of the existence of thousands of planets, despite the fact that very few of them have yet been directly imaged.

An Ideal Planet?

Of the trillion or so orbiting planets in the Milky Way galaxy, only a small fraction – but a pleasingly large number – have been discovered, named, and described. Most of these 5000 or so planets are not conducive to life, but it was always expected, on the basis of knowledge about our own solar system, that the majority of planets would be uninhabitable. So this shouldn't be a cause for despair. Rather, the fact that 50 or so may be habitable – about 1% of the haul – should be a cause for celebration. As a percentage, that's at the low end of the range I delineated in Chapter 4 by using a 'pessimistic' 1% and an 'optimistic' 10%. But that's to be expected also, given the bias of our detection methods towards finding large Jupiter-like planets as opposed to small Earth-like ones. I suspect we'll eventually find that the percentage of potentially habitable planets is closer to 10% of the total, though of course much depends on the definition of 'potentially'.

I'm not going to give specific information for 50 planets; rather, in the next section, I'm going to drop another order of magnitude, and focus first on just five of them, and then on just two. The basis for my selection of these few planets to discuss will be the extent to which they satisfy six criteria that can be thought of as collectively defining the 'ideal' situation. Those criteria are the focus of attention here. They are of two very different kinds. Most of them relate to maximizing the chances that life exists on a planet; as discussed earlier, the pragmatic approach is to think in terms of broadly Earth-like life. But the final criterion relates to ease of study, and hence likelihood of coming up with evidence for life, from an Earth-based perspective. Here they are:

1   A planet should be rocky rather than gaseous. Unfortunately, whether it's one or the other is often not known for sure.

2   It should be orbiting in the habitable zone of its host star, preferably not too close to either the inner or outer edge of this zone, given the difficulty of specifying these exactly.

3   It should be orbiting a middle-mass star, not one of the huge ones with fleeting lives (classes O, B, and A), or a red dwarf (class M) that causes habitable zone planets to become tidally locked. So we're looking for a host star that's similar to the Sun (class G), or one that belongs to either of the flanking classes (F and K).

4   It's probably best to avoid the complications of binary and multiple-star systems, and also stars that are ultra-active, being subject to frequent emission of particularly large flares. That's not to say that life couldn't exist in such situations, but rather that with lots of systems to choose from, these kinds are best avoided.

5   The planet (and star) should not be so young that life is unlikely to have got started in earnest yet, however perfect is its situation for future life. A pragmatic choice, based on what we know about evolution on Earth, is to avoid focusing on planets that are less than about two billion years old. That figure relates to detecting photosynthetic life, and is based on

Earth's Great Oxygenation Event. The threshold age extends, of course, if we're looking for intelligent life. In that case, it should probably be more like four billion years.

6   The closer an exoplanetary system is, the easier it is to study. Since even the closest exoplanets are much harder to study than the planets of our solar system, it makes sense not to amplify what are already considerable difficulties besetting our attempts to do things like analyse exoplanetary atmospheres. Exoplanets within 10 light years would be ideal, though there aren't many of them; those within 100 light years should 'doable'; the further beyond this, the more difficult things become.

The trouble with choosing planets on the basis of these six criteria is that we don't always have the correct information. I already noted that it's not always clear whether an exoplanet is gaseous or rocky. Also, sometimes, stars that started off looking like singletons end up being recognized to be binary pairs. And the age of a system is not always known, or is known approximately but with a very wide margin of error. Given these problems, the best way to proceed is to use what information we have, while acknowledging the gaps.

The Most Promising Planets So Far

Let's now look at five potentially habitable exoplanets that score well on our six criteria. As far as I'm aware, at the time of writing, there's no known exoplanet that gets a perfect score, in that it completely satisfies all of them. But a few cases come close. Of these, I've chosen the following: Kepler-442 b, Kepler-452 b, Proxima b, Tau Ceti f, and TOI-700 d. The naming of the first two follows the telescope-star-planet system that I outlined above. Proxima refers to the nearest star to Earth, Proxima Centauri, which is part of the Alpha Centauri triple-star system. Tau Ceti is one particular star in the constellation of Cetus (the whale), designated not by a number this time but by the Greek latter tau. TOI-700 refers to the 700th 'TESS Object of Interest'. In all cases, the basis for the planet's individual letter is the same as before, reflecting the order of the planets from the host star and/or the order of their discovery. Figure 6.1 gives a summary of the extent to which these five planets meet our various criteria of being both potentially habitable (e.g. in the right zone) and potentially study-able (not too far away from us).

Criterion	K442 b	K452 b	Prox b	TauC	T700 d
1) Rocky planet?	✓?	✓?	✓?	?	✓?
2) In habitable zone?	✔	✔	✔	✔	✔
3) Host star FGK?	✔	✔	✘	✔	✘
4) Non-binary, non flare?	✔	✔	✘	✔	✔
5) Over 2 BY old?	✓?	✔	?	✘	✓?
6) Within 100 LY?	✘	✘	✔	✔	✘?

Figure 6.1   Five exoplanets judged against six key criteria. Most of the criteria relate to the features of the planet itself, its host star, or a combination of the two. However, the last one – distance from Earth – relates to ease of study. Each cell of the table is given a symbol indicating how each planet fares in relation to each particular criterion. In between a solid tick (criterion satisfied) and a solid cross (not satisfied) are the intermediate cases: probably satisfied, unknown, and probably not satisfied. K, Kepler; Prox, Proxima Centauri; TauC, Tau Ceti f; T, TESS; FGK, star classes (as explained in the text and in Figure 2.3); BY, billions of years; LY, light years.

We'll now focus in on two of our five shortlisted planets in particular: Kepler-452 b and Proxima b. These represent an interesting contrast. The former looks particularly promising in terms of habitability, but is disappointingly distant. The latter is less promising, largely because of the nature of its host star, but it has the distinction of being our nearest neighbour.

When Kepler-452 b was discovered, in 2015, it was heralded by some as 'Earth Mark 2', though this label was perhaps a bit premature. It orbits a star that's in the same class as the Sun. Its year is very like ours, at 385 days. It's considerably bigger than Earth – about five times more massive – which puts it in the category of so-called 'super-Earths', namely planets with masses between about double and ten times that of our home planet. Its orbit is in the habitable zone, at a very similar distance to its host star as we are to ours – 1.05 compared with 1.00 AU. The Kepler-452 system is thought to be about 6 billion years old. If this is correct, it's about 1.5 billion years older than our solar system, which might mean it may be coming close to the point at which any oceans will be boiled off by its gradually more energetic host star – but it shouldn't have reached that point yet. All this sounds very promising, but now for the crunch, namely its distance from us, which is estimated to be a little over 1800 light years.

Proxima b has a pattern of pros and cons that's complementary to Kepler-452 b. On the plus side, at a mere four and a quarter light years away, Proxima Centauri is the nearest planetary system to our own. The downside is that this star isn't very conducive to the habitability of planets orbiting it. Proxima b *is* in its habitable zone. However, its host star is a red dwarf, a flare star, and a member of a three-star system. None of these facts rules out the possibility of Proxima b having life, but they don't help. Most problematic, perhaps, is the likelihood that the planet is tidally locked to its host star, with one side permanently irradiated, the other dark.

The closeness of Proxima b puts it in a unique position from the perspective of possible future exploration. With advances in technology, a spacecraft might be able to reach it within a few decades. Indeed, there's a plan – Breakthrough Starshot – to send tiny 'nano-craft', powered by lasers, to the Alpha Centauri system. This is another of the Breakthrough Initiatives, which we first encountered in the previous chapter with the plan for a mission to Enceladus. The aim is to develop the relevant technology to the point where a speed approaching 20% of that of light becomes possible. This would lead to a journey time of between two and three decades, depending on exactly how fast the craft could be propelled. Naturally, radio messages from the vicinity of Proxima b would take four and a quarter years to reach Earth. So the combined time from launch to receipt of information about Proxima b from its neighbourhood could be as low as about 25 years. I suspect this is overly optimistic, but there's no harm in aiming high. The idea that we might be able to extend actual physical exploration of planets to other systems than our own is sufficiently exciting that it's worth pursuing despite the considerable challenges.

To conclude this section, where do things stand at present regarding habitable planets in other systems than our own? We've seen that there are now lots of known planets that *may* be habitable, but none that correspond to the hypothetical 'ideal planet' with which we started. However, new exoplanets keep being discovered. The list of possibilities will grow and grow. At the time of writing, TESS is still engaged in its planet-hunting endeavour; and even if it were to cease activity tomorrow there's a huge amount of data from it waiting to be analysed. In the end, we'll have more than enough promising planets with atmospheres we can analyse for evidence of life. That analysis is the subject to which we now turn.

Analysing Atmospheres

The search for extraterrestrial life is essentially a search for two things: bio-signatures and technosignatures. The former are possible signs of metabolizing life, the latter possible signs of intelligent life. We'll look at technosignatures in Chapter 8; here, we'll focus on biosignatures, and in particular on atmospheric information that can be interpreted as such.

The exact way in which biosignatures are defined is important. Here, I'll use the following working definition. A biosignature is a pattern that indicates the possible, probable, or certain existence of life. This use of 'pattern' is very broad. One very concrete kind of three-dimensional pattern is a fossil. Some fossils – for example a well-preserved trilobite – can be interpreted as 'certain life'. If *Perseverance* finds a fossil trilobite embedded in the rock of Jezero crater (it won't!), this would be conclusive evidence of past life on Mars. Other fossils can be interpreted as probable life. This is true, for example, of some stromatolites – say those with a moderate degree of preservation. In fact, given their varying ages and states of preservation, stromatolites can be anywhere on the spectrum from possible to certain. Other 'apparent fossils' don't get beyond suggesting 'possible life'. This is true, for example, of poorly preserved small round structures that might or might not be cells. In Chapter 5, we saw the dubious interpretation of such 'fossils' in a meteorite, and the equally dubious interpretation of Martian spherules as fungal cells or spores.

When it comes to exoplanets, we can't hope to find biosignatures embedded in their rocks. Their distance away from us renders such a prospect impossible – at least for the foreseeable future. But that's not necessarily a problem, because we can look for different kinds of pattern that are suggestive of life. The most promising kind is a pattern of absorption of light by constituents of exoplanetary atmospheres. Examining these patterns takes us to the science of spectroscopy, so I'll need to explain its fundamentals. If you already know these, please skip the next paragraph.

Gases in exoplanetary atmospheres may leave an imprint on the light that reaches us from the system concerned in a way that identifies them – in effect they 'sign' the light. Remember the dual meaning of 'light'. Here, I'm using it in the broad sense of the whole electromagnetic spectrum; some of the gaseous signatures are in non-visible parts of the spectrum, for example the infrared. As noted in Chapter 1, the signature of one gas is different to that of another, because the unique properties of each type of atom or molecule cause a different pattern of absorption of light when plotted against wavelength. Each pattern consists of a series of discrete absorption bands – usually several of them. Thus we can distinguish, for example, the absorption patterns – or signatures – of oxygen, methane, and carbon dioxide. All these gases, and particularly oxygen (next section), are of interest in the search for life. To conduct the necessary spectroscopical observations to detect them, we need to be able to search the appropriate parts of the spectrum. Usually, this means the ultraviolet, visible, and infrared sections. These sections aren't precisely defined, but, in nanometres (nm; 1 nanometre is a billionth of a metre), they are roughly as follows: 100–400 (ultraviolet), 400–750 (visible), and 750 to about a million (infrared). Because the last of these is a very broad section, it's often subdivided into near (N) and far (F) infrared, with the former being near in the sense that it borders the visible band (near-infrared is from about 750 to 1500 nm).

Now imagine a space telescope orbiting above the absorbing gases of the Earth's atmosphere. Its large mirror is pointed towards a planetary system that's edge-on from our perspective, and not too far away – say within 100 light years. A habitable-zone planet transits across the front of the host star. The telescope collects the light during the transit. This light is analysed by an onboard spectrometer, and an absorption band is found at a particular wavelength – say one that corresponds to a known absorption band of carbon dioxide. A reasonable interpretation of this finding is that CO_2 in the planet's atmosphere is responsible. But space isn't empty, and there's a lot of it in between the exoplanetary system and our own. Might some molecular cloud halfway between the two systems be doing the absorbing? The way to test for this is to repeat the light collection and analysis after the planet has finished its transit. If the band is now gone, the planet must have been causing it.

This particular finding would be interesting, but hardly exciting from the perspective of finding life. We know that in our own system both Venus and Mars have atmospheres in which the main component (more than 95%) is carbon dioxide. So we may have detected an exoplanet that's a bit like one of them. Carbon dioxide in a planet's atmosphere shouldn't be regarded as a biosignature. But perhaps oxygen should? The literature on this subject is full of excitement about oxygen as a biosignature, and almost equally full of caveats that we shouldn't assume that oxygen is indicative of life. We'll now probe this issue.

Oxygen as a Biosignature

Let's start with oxygen as an element. It's just a single one out of almost 100 elements that are described as naturally occurring. However, it's one of the most common of them, both on Earth and in the universe at large. In fact, it's the third most common element in the cosmos after hydrogen and helium. It's found in space, in rocks, and in water. As an element, oxygen definitely should not be considered to be a biosignature. What we're interested in from a looking-for-life perspective is something much more specific: gaseous molecular oxygen (O_2) in a planet's atmosphere. Such atmospheric oxygen can be a biosignature, because it can result from an activity of life forms – photosynthesis. But we have to be careful, because it can also arise from other sources. More on that shortly.

How do we detect O_2 spectroscopically? Like many other gases, it has a very definite pattern of light absorption that's restricted to particular wavelengths. In other words, when light passes through oxygen and is then split into its different wavelengths, we see dark bands at particular points, corresponding to those wavelengths that oxygen absorbs. However, in contrast to the simplified example of a single absorption band for a hypothetical gas that was shown in Figure 1.1, oxygen has several. One of these – called the oxygen A-band – is located at just over 760 nm (Figure 6.2). This is close to the boundary between the visible and infrared sections of the spectrum, which is variously stated to be at 700, 750 (the figure I used above), or 780 nm. So, to detect it,

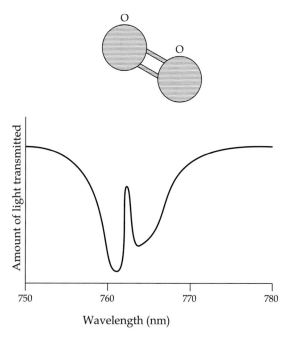

Figure 6.2   One of the tell-tale signatures of oxygen molecules (O_2) on light: the oxygen A-band. This takes the form of a double-dip in the amount of light received at a wavelength of just over 760 nm, which is near the boundary between the visible and infrared parts of the spectrum. This is not the only way that oxygen molecules sign light. There is also a less pronounced B-band at a wavelength of about 690 nm (not shown here), and hence within the visible part of the spectrum. Note that these bands can be shown either as dips (here) or as peaks, depending on whether the vertical axis is the amount of light transmitted through the gas or absorbed by it.

a spectrometer needs to be able to analyse both red and near-infrared wavelengths. The James Webb telescope has such an instrument; it's called NIRSpec (Near Infrared Spectrograph). At the time of writing it hasn't yet been used, but by the time you're reading this page, it hopefully will have been. I look forward to seeing its results in due course.

Before considering what Webb might find when it looks at exoplanetary atmospheres, it's a good idea to consider the atmospheres of solar-system bodies, since these have been characterized in detail. Only one body – Earth – has significant atmospheric oxygen, though even this statement needs to be treated with care, because we need to distinguish between the amount and the proportion of oxygen, as well as the thickness of the atmosphere. Earth has a thick atmosphere – though not nearly as thick as that of Venus – and the proportion of oxygen in it is about 21%. Most of this derives from photosynthesis by living organisms – plants, diatoms, cyanobacteria, and others. The ozone in our atmosphere derives from oxygen, and thus can also be said to be biogenic.

But what of the other planets in our system? Oxygen is absent, or present only in trace quantities, in the atmospheres of Venus and the gas and ice giants. Most listings of the constituents of the atmosphere of Mars just give its main few gases and don't include oxygen; however, it's there, and in more than trace quantities. It occurs at a level of about 0.2%. But the real surprise is Mercury. Most accounts of this planet state that it has no atmosphere. Is that true? Well, yes and no. There *is* a Mercurian atmosphere, but it's incredibly thin compared to our own, and even compared to the rarefied atmosphere of Mars. We've discovered that oxygen is its number one component, and this gas is present both in atomic (O) and molecular (O_2) form. So, in *proportional* terms, oxygen is commoner in the atmosphere of Mercury than in that of Earth, but the actual *amount* of oxygen is tiny.

In contrast to the situation on Earth, the oxygen in the atmospheres of Mercury and Mars isn't biogenic. So in our system there's a link between having a lot of atmospheric oxygen and the existence of life. Is this also true elsewhere? If it is, then it greatly facilitates our search for evidence of biospheres on exoplanets. My guess is that it is indeed true, but we should consider an alternative possibility – that some uninhabited exoplanets have oxygen-rich atmospheres because of rapid non-biogenic production of this gas, for example due to photochemical reactions. Since this possibility can't be ruled out, we need to find some way to diminish the degree to which it complicates the search for life.

Here's one such way. Cast your mind forward to the time when Webb or another space telescope has discovered ten exoplanets with strong oxygen signatures. Now suppose that all ten are in the habitable zones of their host stars, as shown in Figure 6.3 (top panel). Such a finding could be interpreted in three ways. First, the link between having a large amount of atmospheric oxygen and the presence of life is not just restricted to our own system, it's a *general* one. Second, it's just a coincidence, in other words it's simply due to the role of chance in a small sample (in this case ten) and it'll disappear when we know of considerably more planets – say 100 – that show strong oxygen signals. Third, for some as-yet unknown reason, photochemical or other abiotic means of oxygen production tend to be maximized in the habitable

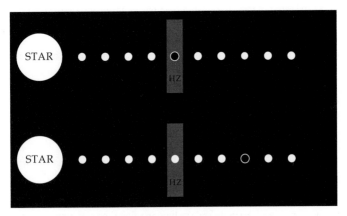

Figure 6.3   Comparison of two possible scenarios for a planetary system in which there are ten planets overall, but just a single one of them in the habitable zone (HZ), and a single one of them from whose atmosphere we detect atmospheric oxygen. Top: the two coincide – in other words, they're the same planet. Bottom: they don't coincide – in other words, they are different planets. If we detect many cases of the former and none of the latter, this pattern would be a pointer to the oxygen being biogenic.

zone. I'll call these the biogenic, chance, and chemical hypotheses, respectively.

The chance hypothesis doesn't have a lot going for it, as a few quick calculations using simplified round numbers will show. Imagine ten systems each with ten planets. In each system, one planet is in the habitable zone, and one shows a strong oxygen signature. What's the probability of these being *the same planet* just by chance? It's one in ten to the tenth power, which works out at one in ten billion. That's easily low enough to write it off as implausible. The chemical hypothesis is harder to test, but I'd say it's also implausible. Another way to put this is that the onus is on its proponents to come up with a credible mechanism for abiotic production of oxygen in the atmosphere to be restricted to planets that are in the habitable zone. This leaves us with the biogenic hypothesis as the most likely explanation of the pattern found.

Of course, nature is often messy, and we might find instead that of the first ten planets that show strong oxygen signals, eight are in the habitable zone (as in the top panel of Figure 6.3) and two are not (as in the bottom panel). Such a pattern is easier to explain by small sample effects than is the 10–0 pattern, though its probability of being a chance result is still very low. Anyhow, whatever the exact pattern in a sample of ten, when we reach our first 100 exoplanets with strong oxygen signals, the reality or otherwise of the pattern should be clear.

One final point needs to be made here concerning the search for biosignatures in exoplanet atmospheres. Oxygen is the most important gas to search for with spectrographic techniques, but it isn't the only one. We can also search for other gases that may be indicative of life. I've already mentioned the possibility of looking for oxygen's derivative, ozone (O_3). This has a much stronger signature than that of O_2, and so it may be easier to detect, even if present at a lower level, as is the case in our own atmosphere. Another gas we can search for is methane (CH_4). In particular, the sustained co-occurrence of significant amounts of oxygen and methane might be considered as suggestive of life, because left to chemical equilibrium in the absence of biogenic input – for example from microbial methanogens – such co-occurrence would be transient. However, in the end oxygen is the most important of all. In her book *Exoplanet Atmospheres*, the Canadian-American planetary scientist Sara

Seager describes oxygen as 'Earth's most robust biosignature gas'. It may well merit this description in the atmospheres of alien worlds too.

How Long Until Discovery?

It's a reasonable guess that discovery of oxygen in the atmosphere of an exoplanet in the habitable zone of its host star will be the first persuasive – albeit not conclusive – evidence of alien life. But when might we expect this to happen? Is it something that some of us will live to see, or is it merely a remote possibility in the very distant future? I started this book with the view expressed by James Green et al. in a 2021 paper in *Nature* that our generation will be the lucky one – a view with which I agree. But on what bases should we be optimistic about finding the first evidence for life within the next couple of decades? There are two of them, as follows.

First, we can look at the rapid pace of discovery to date in the field of exoplanet science in general and exoplanetary atmospheres in particular. In terms of the nature of the planets themselves, we've gone from pulsar planets to hot Jupiters to Earth-like planets in habitable zones within about 20 years. And in terms of the nature of exoplanetary atmospheres, we've gone from knowing almost nothing at the start of the millennium to having information on atmospheric constituents of multiple exoplanets today. The gases that have been discovered so far include hydrogen, helium, carbon monoxide, carbon dioxide, methane, water vapour, and hydrogen cyanide. The first exoplanets subjected to atmospheric analysis were hot Jupiters, but since then research has extended to include super-Earths as well.

Second, looking ahead, we can consider the evolution of the appropriate technology, in particular the spectroscopic devices of current and future space telescopes. We've already noted Webb's capability for spectroscopic analysis in the red and near-infrared regions, where the oxygen A-band is found. It's not the first space telescope to have this capability – an earlier one was Spitzer – but the design of its NIRSpec is greatly enhanced compared with previous versions. And the planned space telescope based on the LUVOIR design, which I mentioned earlier, will have improvements on Webb's capability in terms of observing exoplanetary atmospheres. It will be specifically focused on directly imaging the infrared radiation coming from exoplanets by using some form of starshade to blot out the host star. This can already be done to a degree. Devices called coronagraphs were developed for this purpose to observe the Sun's outer corona, and Webb has these among its arsenal. But the next-generation space telescopes will be even better from this perspective. Also, we shouldn't forget that there is a new generation of very powerful ground-based telescopes in the planning and construction phases, for example the ELT (Extremely Large Telescope) currently being built in Chile and planned to be complete in 2025.

Webb's mission is scheduled to last for at least ten years, and hopefully it will last longer than that. Some earlier space telescopes have considerably exceeded their planned mission duration. With luck it will last until around 2040, and that's about the time when a LUVOIR-based design should be ready to launch. That later scope will probably be operational until about 2050 or 2060. I suspect we'll have our first evidence for biogenic oxygen long before 2060. I'm an optimist, and if asked to guess a date and telescope for this phenomenal finding I'd say around 2030 and Webb. Time will tell whether I'm right in making this optimistic prediction.

7 The Nature of Extraterrestrial Life

Conflicting Hypotheses

While we wait for our first conclusive evidence of life beyond Earth, we can contemplate its possible nature. In particular, we can ask the following question. To what extent should we expect evolution elsewhere to take a similar course to the one it has taken on Earth? That could be described as the key question about the biology, as opposed to the geography, of extraterrestrial life. But the way I've just put it isn't ideal – it's too centred on our home planet as a reference point. Let's try to rephrase it in a Copernican manner, so that Earth doesn't occupy a special place. Here's one such rephrased version. To what extent does evolution follow similar courses on different inhabited planets? Earth is implicit here, but just as one of many inhabited planets, and almost certainly not the first one.

This question about the similarity of evolutionary processes and the life forms they produce on inhabited planets scattered across the observable universe has an important alias: what is the scope of the branch of science that we call biology? The scope of physics is clear: it's universal. Stars burn in the same way as our Sun, however distant they may be. The same is true of chemistry: it also applies everywhere. The way in which two chemical elements interact is dependent only on conditions, not on place. On a distant planet adorned with both continents and oceans, and with a similar atmosphere to Earth's, a sheet of iron left on a beach for a decade will rust just as surely as it does here.

But what about biology? There are two opposing views on this issue. One is that biology is a parochial affair, applicable only to a single planet – Earth. The other is that biology, like physics and chemistry, applies across the universe. I support this latter view. I suspect that life forms everywhere are based on carbon, made of cells, and evolve via Darwinian natural selection. I imagine that parallel environments on different planets produce parallel selection, which in turn produces parallel evolutionary trees, and hence parallel arrays of life forms, including animals and plants. The aim of the present chapter is to flesh out this view of life in the universe, and to show why it makes more sense than the alternative 'parochial biology' view.

Let's give names to these two opposing views, with their alternative emphases on the similarity versus dissimilarity of life between one inhabited planet and another. Perhaps 'universal biology' versus 'parochial biology' would work, but I think we can do better than that. Although I support the former rather than the latter, I shouldn't name them in a way that might prejudice readers in favour of my preferred view. The word 'universal' has positive overtones, while 'parochial' tends to be a pejorative term. Instead, let's use phrases that are more neutral: *parallel life* versus *contrasting life*. Although such a pair of opposing labels might be taken as suggesting two clearly defined hypotheses, it's important to realize that this is not the case. Rather, each of the two labels refers to a bundle of related hypotheses that includes moderate and extreme versions. We'll now consider this point further, first in relation to parallel life, then in relation to contrasting life.

An extreme version of the parallel-life school of thought is adopted by the English palaeontologist Simon Conway Morris in his book *Life's Solution*. His view is neatly summarized by the book's subtitle: *Inevitable Humans in a Lonely Universe*. Conway Morris sees evolutionary processes as inevitably generating humanoid forms. His emphasis is on the possible alternative trajectories that evolution on Earth might have taken, but the implications for other planets are clear. The basis of his argument is the commonness of convergent evolution, which is the evolutionary production of similar organismic forms from different starting points, an example being the evolution of the birds we call swifts and swallows. Other authors who have championed the importance of convergent evolution in arguing for the probable existence of similar life forms on different planets are David Darling, in his book *Life Everywhere*, and George McGhee, in *Convergent Evolution on Earth: Lessons for the Search for Extraterrestrial Life*.

Other approaches to the parallel-life view have been adopted by several authors. Charles Cockell, in *The Equations of Life*, approaches it from the perspective of physics. In *The Cosmic Zoo*, Dirk Schulze-Makuch and William Bains examine it from the viewpoint of chemistry and cell biology. Zoologist Arik Kershenbaum uses animal behaviour, including sociality and communication, as his starting point in *The Zoologist's Guide to the Galaxy*. Here, I adopt a multifaceted approach, centred on evolutionary biology, as I did in my 2020 book *The Biological Universe*.

The contrasting-life school of thought also comes in various versions. In the most extreme of these, in which non-carbon-based life is considered possible or even likely, the idea of parallel life based on carbon being the norm is written off as carbon chauvinism. However, not many scientists support this version. A less extreme view is the 'rare Earth' hypothesis that the American scientists Peter Ward and Donald Brownlee put forward in their book *Rare Earth: Why Complex Life Is Uncommon in the Universe*. These authors accept that while carbon-based microbial life may be common, multicellular life forms such as animals and flowering plants are rare. Then there's the view of James Trefil and Michael Summers, espoused in their book *Imagined Life*. Although they accept some aspects of the parallel-life approach, including the likelihood of both carbon-based life and multicellular life on habitable exoplanets, they also consider ideas that emphasise contrasting life. For example, they imagine a world where intelligent life forms are conversing about the possibility of surface life – from their base, which is a university located near a deep ocean hydrothermal vent on some distant exoplanet.

There are good reasons to reject all these scenarios that are based on the contrasting-life perspective. As noted in Chapter 3, elements other than carbon lack the ability to form macromolecules that can rival DNA in their sequence specificity, and hence in their informational capacity. The idea that there is some kind of 'probability barrier' between unicellular and multicellular life, rendering the former common and the latter vanishingly rare, is hard to defend. This is especially true because multicellularity has evolved on Earth many times, indicating that this is a relatively easy evolutionary transition. And the idea that a technological civilization can be produced in the deep oceans shouldn't be taken seriously unless its proponents can come up with either some way for metals to be produced and used there, or some credible substitute for metals that can form the basis of the technology concerned.

What I'm saying here is that a few basic observations and some simple logical arguments point us in the direction of parallel life rather than contrasting life on different inhabited planets. But, as I've said, parallel life is a range of possibilities, not a single clear-cut hypothesis. It comes in various forms. The most extreme form, in which the evolution of humans is somehow inevitable, is hard to defend. In the hands of its main proponent, Simon Conway Morris, the motivation for it is at least partly theological, which is not a good basis for doing science. Here, I'll make the case for a more moderate version of parallel life, starting with 'parallel basics'.

Parallel Basics and Autospermia

In Chapter 3, I divided the attributes of life on Earth into 'basics' and 'elaborations'. The foundation of this distinction was whether the attributes concerned have existed since evolution began – the basics – or whether they have changed dramatically as evolution has proceeded. The basics include the centrality of carbon, the aqueous nature of metabolism, the use of nucleic acids as information-storage molecules, and the adoption of the cell as the basic constructional unit, whether it exists on its own (unicellular life) or as a 'brick' used to build bigger bodies. In contrast, many attributes of those bigger bodies are elaborations of the basics, and they can change markedly in the course of evolution – for example in the invasion of the land by both plants and animals.

I've already made the case for parallel basics, by arguing that contrasting basics – for example silicon instead of carbon – don't work. Thus the argument is that in the early days of a planet, carbon-based life gets going while chemically contrasting life doesn't, as a consequence of the difference between organic and inorganic chemistry; and that this applies regardless of whether the planet concerned is Earth or some distant exoplanet basking in the sunlight of an alien star. However, it's important to realize that there could conceivably be a very different reason why life might have parallel basics on different planets – panspermia. This is the idea that life on Earth was seeded with spores from elsewhere, or, to generalize that notion, life on any one inhabited planet started with the influx of spores from another.

The panspermia hypothesis is usually attributed to the Swedish scientist Svante Arrhenius (1859–1927), though its roots go back further than his day. It was promoted by the British astronomers Fred Hoyle and Chandra Wickramasinghe in the mid-twentieth century. Arrhenius was a Nobel laureate, Hoyle was a Fellow of the UK's Royal Society, and Wickramasinghe was awarded the Vidya Jyothi – his native Sri Lanka's highest scientific honour – so these were clearly esteemed mainstream scientists. Yet panspermia is generally not regarded as a mainstream theory, and I think rightly so. It stretches credulity to the limit in terms of the conditions it requires life forms to survive – basically, those of deep space.

An important distinction needs to be made here – that between the atoms and molecules that are needed for life coming from space on the one hand, and ready-made organisms in the form of spores or other resistant life-cycle stages coming from space on the other (Figure 7.1). The former clearly happens, because planets form from proto-planetary discs, which themselves form from clouds of gas and dust in space, and the elements in these clouds were manufactured in earlier stars. But the latter – living spores coming from space – almost certainly doesn't happen, because the distances between one planet and another are too great, and thus the travel time between them too long, for life to be sustained.

When panspermia was first proposed, we knew nothing of the ability of life forms to survive in space. That has now changed, as various species of Earthly life have been taken into space on various missions. The ones that are relevant here are those where the organisms concerned were subjected to the conditions of space itself rather than being cosseted in the spacecraft's interior. For example, those little invertebrates called tardigrades, which we encountered in Chapter 1 (refer back to Figure 1.2), have been taken into space and survived. The first mission of this kind – a joint one between the Russian and European space agencies – took place in 2007. In this and subsequent missions, some of the tardigrades survived, and this is the result that made headlines. However, some did not – a fact that's often swept under the proverbial carpet. Also, their journey was, by the standards of interplanetary travel, a local one. You can't compare being a short distance above Earth's atmosphere to making the journey to Mars, even less so to the journey from one planetary system to another.

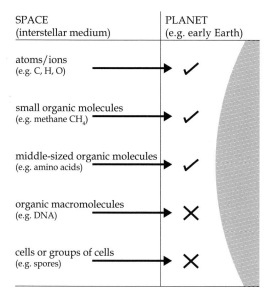

SPACE
(interstellar medium)

PLANET
(e.g. early Earth)

atoms/ions
(e.g. C, H, O) ✓

small organic molecules
(e.g. methane CH_4) ✓

middle-sized organic molecules
(e.g. amino acids) ✓

organic macromolecules
(e.g. DNA) ✗

cells or groups of cells
(e.g. spores) ✗

Figure 7.1   What comes from space and forms the basis of life on Earth, or indeed on any other planets? Atoms/ions for sure, but cells almost certainly not. Here, these and three intermediate possibilities are pictured. The illustration represents the hypothesis that everything up to medium-size organic molecules are found in space and on uninhabited bodies, while sequence-specific macromolecules such as DNA and proteins are not. Another way of stating this is that DNA and proteins are necessarily biogenic, while smaller molecules can be made by purely chemical processes.

When we consider what an interplanetary or interstellar journey would entail for an Earthly life form, the prospects of survival rapidly diminish to zero. Even at the very start, the odds are against it: instead of being launched into space by a rocket, spores or other life forms have to float up through the atmosphere. Once outside it, they are subject to the Sun's full range of electromagnetic radiation, from the most damaging parts of which all organisms on Earth are protected. Despite lacking any form of propulsion, they then have to drift through interplanetary space in the direction of another potential world. Even a powered trip to Mars takes years – how many depends on the positions of the planets in their orbits. A non-powered trip might take millennia or more. And interstellar trips would take millions of years. Approaching an exoplanet, a wandering spore would be once again blasted by stellar radiation – this time from a different star. No Earthly life form could survive such a trip, and probably no extraterrestrial life form either. In contrast, the issue of staying alive doesn't feature with space-travelling atoms and molecules.

The opposing hypothesis to panspermia needs a name. For Earth, it can be called terraspermia – life on Earth originated on Earth. But that can be generalized to autospermia – life on a planet originated on that planet. This potentially applies throughout the observable universe. But can we envisage exceptions? Yes, and of two very different kinds: first, where two or more habitable planets occur very close together in the same system, as in the case of TRAPPIST-1; and second, where intelligent life has colonized one or more planets close to its home one, as may be the case in the future with human bases on Mars. I suspect that the second of these is a reality somewhere, but that the first is not.

In conclusion, there are two possible ways in which parallel basics of life might arise on different planets: panspermia on the one hand, and a version of autospermia involving parallel origins of life on the other. While the former can't be ruled out, the latter is much more probable. But what will happen next? Even if life starts off in parallel ways on different planets, it needn't evolve in parallel ways thereafter. However, there are many reasons why it might indeed do so, some of which we'll now explore.

Parallel Natural Selection

One of the fundamental distinctions between the realms of life and non-life is the occurrence of Darwinian natural selection in the former but not in the latter. As soon as life arose on Earth, it became subject to selection, and changed as a result – the process is of course ongoing today. Although as yet we have no information about evolution on other planets, it's reasonable to expect the same to apply. The American biologist George C. Williams began his 1992 book *Natural Selection* with a statement of what he called a philosophical position: the view that selection would apply on all inhabited planets wherever they may be in the universe. I share his position, and although I agree it's a philosophical one, I think it has a scientific basis too, as follows.

The features required for natural selection overlap with those involved in the definition of life. Natural selection requires three features in order to happen: variation, reproduction, and inheritance. Life is defined by three features: reproduction, inheritance, and metabolism. So reproduction and inheritance feature in both cases. Variation isn't included in the definition of life because it doesn't help to distinguish life from non-life. Variation is ubiquitous, regardless of whether the entities concerned are stars, planets, and moons on the one hand, or bacteria, plants, and animals on the other. Non-life is *not* characterized by metabolism, but arguably metabolism isn't needed for natural selection. After all, those non-metabolizing inhabitants of the grey area between life and non-life that we call viruses evolve by Darwinian selection.

Charles Darwin's brilliant exposition of natural selection in *The Origin of Species* was full of detail about life forms on Earth. But the logic of his argument was not restricted to our home planet. Rather, stripped to its bare essentials, his argument was one that applies to any inhabited body – planet or moon – across the entire universe. Naturally, the course of evolution wouldn't be expected to be identical between one planet and another, because their environmental conditions are inevitably different to some degree. But equally, evolution may follow broadly similar paths if the environmental differences are quantitative rather than qualitative, as seems likely – for example water always existing in the habitable zone, but varying in amount and distribution.

Here's a scenario for natural selection on an exoplanet. Life originates as unicellular organisms in aquatic environments. As it begins to spread from its point of origin, it encounters different ecological conditions. As it does so, natural selection alters it in ways that adapt it to the conditions concerned. Lineages diverge based on the selectively driven differences that accumulate between them. Some of the differences are in the chemical processes that go on in the cells concerned, which will be largely unrecorded in the planet's fossil record. But other differences will arise that leave traces for observers to detect. Key among these will be increases in organismic size, achieved in most cases by the evolution of multicellularity – a process that we've seen has happened many times on Earth. Once in the realm of the multicellular, the evolution of organismic shape and size are effectively liberated into a new dimension, and multifarious forms, including some very large ones, emerge.

Studies of evolution on Earth often distinguish between pattern and process. Birds evolved from a particular group of dinosaurs – that's a pattern, which can be represented by a particular evolutionary tree diagram. The process that drove it included natural selection for feathers and flight. Pattern and process are connected, but in a loose way. One particular pattern of relatedness doesn't necessarily imply a particular form of natural selection; and one particular form of selection doesn't imply a particular pattern. For example, the dinosaur–bird link is true regardless of the selective reason that feathers were favoured – perhaps for thermoregulation first and for flight later. Also, applied to a different starting lineage, selection for flight can produce the very different wings of bats.

This distinction between pattern and process is likely to be true elsewhere also, with the two being connected, but in a rather loose way. So an interesting question arises: if natural selection operates in broadly parallel ways on different planets, might this mean that they end up with broadly parallel evolutionary trees? This is a distinct possibility, though it's necessary to emphasize that 'broadly' covers a range of degrees of similarity.

Parallel Trees of Life

In relation to evolution on Earth, there's sometimes debate about whether 'convergent evolution' or 'parallel evolution' is the better label for situations in which similar forms arise in widely separated lineages – for example placental and marsupial moles. Personally, I see the mole example as parallel, because in both cases evolution produced an underground (U) descendant from an above-ground (A) ancestor, so it's a process of U → A in both cases, in other words, parallelism – but some authors call it convergence. However, from an astrobiological perspective this is really a storm in a teacup. What's far more important when it comes to making comparisons between the overall trees of life on different planets is that the meanings of terms like parallelism and convergence change when they're applied at the whole-tree level. That's what we need to consider now. We'll start by looking at the tree of life on Earth, and then consider possible alien counterparts.

As we saw in Chapter 3, a billion years ago on Earth, there were many kinds of microbes. From particular groups of these arose 'macrobes' – multicellular life forms of one sort or another. There have been many origins of multicellularity in the history of life on Earth, as we've seen (refer back to Figure 3.3). This shouldn't surprise us. Microbes often associate together in various ways – they're not purist single-celled loners with no interest in socializing. All that needs to happen in any particular lineage that goes multicellular is for the 'sociality' to become obligate rather than optional.

Animals arose from a group of microbes called 'collar-whips' (an unofficial translation of the official term choanoflagellates). This name is used because, although each cell is roundish for the most part, it has a collar-like structure at one end, with a hairlike whip (or flagellum) protruding out of it. Sponges are among the simplest animals on Earth today, and one of their main constituent cell-types is rather similar to the collar-whips. The evolution of a particular ancient lineage from intermittent aggregations of individual collar-whip cells to a system of programmed development into proto-sponges that repeats itself every generation was the origin of everything, from an animal perspective. The whole animal kingdom, with its millions of species, both extinct and extant from our present-day vantage point, evolved from that ancient, sponge-like, primordial animal.

However, we can also look at this event from a different, and broader, perspective. Usually in evolution, when a brand-new kind of creature originates, its ancestor continues to live alongside it, rather than going extinct. This is just as true of the origin of animals, in the form of continuing unicellular collar-whips, as it is in later divergences, such as the continuation of jawless fish (e.g. lampreys) after the origin of vertebrate jaws, or the continuation of fish in general after one of their lineages diverged in form as it adapted to life on land. What this means is that most *origins* (from one perspective) are also *divergences* (from another). Even the origin of the birds was a divergence, because large land-based dinosaurs coexisted with the early birds until the asteroid-induced demise of the former about 66 million years ago. Whether we choose to look at evolution as a series of origins or a series of divergences doesn't matter – sometimes one view seems more appropriate, sometimes another.

An evolutionary tree such as that of life on Earth is a multiplicity of interconnected divergences. Think again about comparing marsupials with placentals, but this time more broadly than just the mole lineages mentioned above. Both of these large groups of mammals have diversified into a wide range of forms, adapted to a wide range of habitats and niches. The marsupial radiation produced hundreds of species, the placental one thousands. Both radiations gave rise to ground-dwellers (cows and kangaroos), subterranean tunnellers (those moles), aquatic forms (dolphins and the yapok or water-opossum), and gliders (colugos and sugar gliders). There are no powered fliers in the marsupial tree – i.e. no marsupial bats – so the two radiations aren't identical. But we'd hardly expect that. Evolution is a messy process with no eye to the future. Similarity at a detailed level would be odd, given the dependence of the process on chance events. However, the two big mammal groups can be thought of as being *broadly parallel divergent radiations* (note the combination of 'parallel' and 'divergent' here) or, more succinctly, parallel trees (Figure 7.2). I now want to persuade you that this is what we should expect

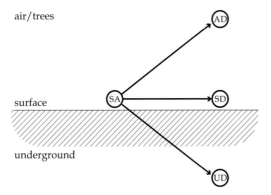

Figure 7.2   Two broadly parallel evolutionary radiations on Earth, those of the placental and marsupial mammals, both fit the simplified model shown. From a starting point of a surface-dwelling ancestor (SA), both trees produced surface, underground, and aerial descendants (SD, UD, and AD), though while the placentals evolved both active (bats) and passive (colugos) aerial forms, the marsupials only evolved the latter (sugar gliders). This picture of broadly parallel radiations may also apply in comparisons between the tree of life on Earth and that of an inhabited exoplanet.

to find when we're able to compare the whole tree of life on Earth with its counterpart on a distant planet.

Let's imagine a combined journey in space and time to a particular Earth-like planet basking in the habitable zone of a Sun-like star that's a few hundred light years from home. We arrive there when the planet is about half a billion years old. Life has started, but is still in its infancy. The first rickety proto-cells that were prone to falling apart have been replaced by cells of better coherence. The microbial tree has begun to radiate. There are many different kinds of unicellular life forms, but nothing big. Large life forms may lurk in the evolutionary future of this young planet, but there are none in 'the present' – a term that becomes hard to define if time travel is possible, but we'll take it to be our time of arrival at the planet concerned.

Now imagine a journey into this planet's future. What happens as life continues to radiate? Will yet more kinds of microbes appear? Almost certainly. Will large multicellular life forms originate? Very probably. They've originated many times on Earth, so why not on another similar planet, given a potential evolutionary future of billions of years. Will one of these origins result in a large branch of life – a kingdom if you like – consisting of generally mobile creatures that make their living by eating other organisms? This is by no means assured, but I'd say that the likelihood of alien animals is high.

Now the questions get harder. Will vertebrates eventually evolve on our study-planet as we visit it at intervals spaced out by – say – half-billion-year leaps in time? Possibly. If animals evolve past a certain size, the evolution of some kind of skeleton is likely, but there may be planets on which armoured animals like Earth's arthropods get such an early and comprehensive hold on all available habitats that bones never evolve at all. If there are no vertebrates, there will be no mammals; if there are no mammals, there will be no primates; and if there are no primates there will be no humans. We are by no means inevitable outcomes of an animal evolutionary tree.

What we're beginning to glimpse here is a fundamental property of evolutionary trees or radiations: early divergences constrain what's possible in later ones. If the probability of animals arising is P, then the probability of vertebrates arising is less than P, and the probability of any particular subgroup of vertebrates arising is less again (Figure 7.3). Some important early divergences

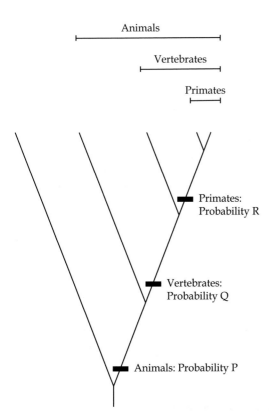

Figure 7.3 A general principle regarding the degree to which the evolutionary trees on different planets might run in parallel to each other. Suppose that the probability of a lineage leading to animals arising is P, then the probability of a lineage leading to animals with bones (vertebrates) arising is Q, where $Q < P$, and the probability of a lineage leading to primates arising is R, where $R < Q$. This is a logically inescapable conclusion about trees that are entirely divergent. If convergence can also happen, then it's not logically necessary, but it might be true nevertheless. For example, it might not be possible to have an arthropod version of a primate.

almost certainly run in parallel between one tree and another, but as time goes on and evolution takes some routes rather than others on each planet, the divergences themselves may diverge! No two trees will be the same in detail.

The reasons for evolution taking certain routes as opposed to others on any particular planet will be of two main types. They could be called – to borrow from Jacques Monod's famous book title – 'chance' and 'necessity'. Alternatively, they could be called 'contingent historical accident' and 'repeatable natural selection'. This latter, more verbose, rendering of the contrasting types of influence on evolution links with the age-old debate about the relative importance of deterministic and stochastic factors. This debate was given an injection of extra 'oomph' by Stephen Jay Gould in his conjectures about the possible role of historical accidents in the Cambrian explosion of animal life on Earth – conjectures that some biologists loved and others hated. Interestingly, despite the intensity of emotions on their relative importance, in terms of constraining possible future creatures it might not matter what drives evolution in certain directions – chance, necessity, or a combination of the two – with the last of these three possibilities almost certainly corresponding to reality.

Consideration of possible evolutionary trees that could arise on exoplanets, and the degree to which they might run parallel to that of Earth, constitutes a fascinating thought experiment. However, it's too large and open-ended a subject to be discussed holistically here. Given limited space in a book of this kind, I must be selective. The basis of my selection, as earlier, is to link in with practical considerations of searching for extraterrestrial life. Remember that we search for both biosignatures, notably oxygen, and technosignatures, notably radio signals. So the evolution of photosynthesis and intelligence are the two trends in the evolutionary trees of other planets that are most relevant to the search for life.

Talking About Selection

In the next two sections I discuss the ways in which natural selection might act on alien organisms in parallel to the ways that it has acted on Earthly ones, with particular reference to features associated with photosynthesis in plants and intelligence in animals. In these discussions, I use a particular form of words that's a kind of scientific shorthand. It allows reasonably brief statements, but that facility comes at a cost, which I need to expose and explain. What I'm referring to specifically here is the use of phrases like 'what selection is for' and 'the target of selection'. You might have noticed that I already used the phrase 'selection for flight' a few pages back; another way of saying this is that flight ability is selection's 'target'. In reality, however, natural selection is not 'for' anything, and it has no 'target'. The idea of a targeted character, such as improved flight ability, or more efficient photosynthesis, or enhanced intelligence, is something that we use to make sense of a complex situation, and to focus our attention on one particular facet of it.

The kind of real selective scenario that fits this terminology best is that which applies in the case of unusually strong selection from a particular environmental source. A good example is the evolution of antibiotic resistance in bacteria. Here, a sudden influx of a lethal chemical causes huge mortality – usually more than 99%. In being such an effective agent of mortality it is also an effective agent of selection: it can drive very rapid evolution of resistance in bacterial populations as the few naturally resistant bacteria proliferate exponentially. In extreme cases such as this, the kind of terminology I'm now focusing attention on is – arguably – justified. The antibiotic here is the selective agent. It produces what we can call a selective pressure on the population. Selection is clearly *for* ability to survive the antibiotic. So this ability could be described as the 'target' of selection.

But most real selective scenarios are not of this kind. Rather, they are situations in which there are many different sources of mortality, and many different organismic characters that have a bearing on survival. It may be a case of a population of finches recently arrived at a new island in the Galapagos archipelago where conditions are different from those in its previous home; or it may be a proto-human recently descended from the forests of Africa. In both cases, selection originates from multiple environmental sources and affects multiple characteristics of the organisms exposed to the new environment. We may choose to focus on one of them – for example finch beaks or human intelligence – but we acknowledge that there are others too. In this sort of scenario, the use of terms like 'what selection is for' or 'target of selection' are harder to defend. But the most important thing to keep in mind – whatever terms are used – is the complexity of the situation and our use of various simplifying techniques in order to facilitate study.

Although my main aim in this section is to point out a terminological concern, it's worth briefly noting a scientific concern too. In those cases where selective scenarios involve multiple organismic characteristics, a complexity arises because of the different ways in which characters may be co-inherited. At the simple end of the spectrum of possibilities are cases in which two selectively favoured traits of an organism are both the direct results of a single gene. This is rare. More often, the genetic basis of pairs and groups of traits is complex, and results from the combined actions of many genes, with partial overlap between the suite of them that affects one trait and the suite that affects another. Not only that, but many traits are only partially heritable; they are also determined in part by direct effects of the environment.

So, starting from a simple unidimensional notion of 'what selection is for', we end up with a complex multidimensional picture of how organisms with many characters fare in environments with many factors that affect survival and reproduction; and how the offspring of these organisms are affected via multifactorial inheritance. Given this situation, the use of a simplifying terminology can be forgiven – but only as long as we keep in mind its limitations. I hope you'll keep them firmly in mind as we consider selective scenarios involving photosynthesis and intelligence in the next two sections.

Harvesting Light

First, then, the evolution of photosynthesis, and, associated with that, the evolution in body form of photosynthetic organisms. We saw the major steps of this evolutionary process on Earth in Chapter 3; here's a brief recap. Oxygen-producing photosynthesis in cyanobacteria arose early, sometime between 3.5 and 2.5 billion years ago. It was associated with the Great Oxygenation Event, in which oxygen began to accumulate in Earth's atmosphere. Cyanobacteria have continued to prosper right through to the present day. The plant kingdom in the broadest sense – including the red and green algae – arose from an early unicell into which cyanobacteria became incorporated via endosymbiosis. Land plants evolved from green algae about half a billion years ago, and diversified rapidly. Tree growth forms evolved many times in various groups, especially among the angio-sperms (flowering plants). However, the angiosperms have a relatively recent evolutionary origin – about 150 million years ago. Separately from the plant kingdom, photosynthesis evolved in heterokonts, including diatoms and brown algae, by a different endosymbiotic event.

So the question now is to what extent equivalents of these evolutionary steps might have happened on other inhabited planets that are of an equivalent age to Earth. To say that this is a big question is an understatement *par excellence*. How can we best approach it? Perhaps the best way is to consider the possible selective pressures on alien organisms, and to work on the basis that selection will usually prevail. In other words, an evolutionary process based on the accidental generation of variation and its sifting by natural selection will usually succeed in finding routes to the solution of problems, even though it will often take a long time to do so.

However, note the 'usually' in the previous sentence. There are some so-called 'universal constraints' on what selection on Earth can do, and, true to their name, these may apply across all inhabited planets. A good example is evolution's apparent inability to produce wheels. This was highlighted by Stephen Jay Gould in his essay 'Kingdoms without wheels', in his 1983 collection *Hen's Teeth and Horse's Toes*. Humans make many mobile machines, and most of them have wheels: cars, trains, planes, and lots of others. Even those that don't have wheels on the outside often have internal wheels, for example those in the engines of ships. Evolution has also made many mobile machines – animals. Yet none of these have wheels. The reason for this is almost certainly linked to the incompatibility of living tissue and the system of disconnected parts that's required for a wheel to spin freely. This constraint urges caution in assuming that natural selection is all-powerful – it's not. But it *is* good at finding solutions to problems. Cheetahs provide a good example of how the problem of needing to catch rapidly moving prey can be solved without recourse to wheels.

Now let's consider the selective pressures that are likely to apply to early life forms on other planets. It may be the case that when the first life forms arise on a planet, they typically acquire their energy by chemosynthesis. But, for life on the surface of a planet orbiting a host star, the amount of incoming stellar energy will outweigh that available from chemosynthesis by many orders of magnitude. Thus any variant life form that can harness this abundant source of light energy is likely to be favoured by selection. This might lead to an evolutionary tree of photosynthetic microbes, like Earth's cyanobacteria, or to a tree of large photosynthetic life forms, like Earth's land plants. Any particular inhabited planet might have both of these. Indeed, it might have *more than one* of both of them. On Earth, the two groups of large multicellular photosynthesizing organisms that we call plants and brown algae are very different in their taxonomic status and number of species – one is considered to be a kingdom, the other merely a class. But on another world there could easily be two groups of large photosynthesizers of approximately equal status – two plant kingdoms, if you like.

We'll now restrict our attention to large photosynthetic life forms on land. Again, there's no reason why Earth's single radiation of such forms shouldn't be replaced by two or more such radiations on other planets. Anyhow, regardless of this, certain selective pressures that transcend individual planets can reasonably be anticipated. One of these is selection for enhanced light collection, which leads to the evolution of broad flat leaves. As on Earth, the balance of selective pressures will change from one environment to another. There may be alien versions of Earth's effectively leafless cacti on parts of a planet's surface where selection to conserve water counteracts selection for maximizing light collection. But, as here on Earth, these should be exceptions to 'the general rule of leaves'.

Once there are plants on land with leaves, there will be competition between them. This will result in selection for increased height. Thus the evolution of woody plants, including bushes and trees, can be anticipated. An inhabited planet with a significant area of emergent land is likely to have trees – at least if it's old enough for the evolutionary processes leading to trees to have taken place. This is an important aspect of the evolution of photosynthetic life. But it may also provide a link with the evolution of intelligent life. We humans are a result of the ascent of our distant ancestors into the trees, followed by the descent of our less distant ancestors back to the ground. On a planet without trees – a 'water world' with a global ocean, for example – there may be intelligent animals, but they are unlikely to be humanoids.

Becoming Intelligent

As we now turn from photosynthesis to intelligence, the same overall strategy applies: we approach the problem from the perspective of the general efficacy of natural selection, while being cautious not to see selection as all-powerful. However, there's an important difference between the evolutionary process that led to complex life forms that photosynthesize, on the one hand, and that which led to complex life forms that are intelligent, on the other. This concerns what's being 'selected for' (recall the philosophical difficulties associated with this phrase!), and whether this has changed over evolutionary time. Arguably, what's been selected for in the evolution of photosynthesis has remained the same over billions of years: the ability to acquire energy from light, the efficiency with which this is done in general, and the ability to adjust the process to different physical environments (e.g. land versus water) and to different amounts of competition from other photosynthesizers.

This continuity of what's being selected for doesn't apply to the evolutionary process that propelled one evolutionary lineage from a small marine sponge-like creature to *Homo sapiens*. This can be clearly seen if we recap the six steps involved, from the account given in Chapter 3. These were: the origin of a nervous system, the advent of bilateral symmetry, cephalization, manipulative appendages, sociality, and complex language.

In the earliest of the steps in the chain that led to intelligence, selection was not for intelligence at all. Rather, it was for movement. This sort of selection was responsible for the evolution of animals with muscles and nerves from simple early forms that had neither of these features. It was also responsible for the evolution of bilateral symmetry from the earliest animal body forms that were asymmetric or radially symmetrical. Arguably, in the process of cephalization, intelligence or something akin to it *was* being selected for, but a very limited form of it for most of the time. More cephalized animals were better able to coordinate patterns of movement to survive in unpredictable environments – and all environments show degrees of unpredictability – but they weren't capable of much in the way of learning.

The evolution of appendages also involved a shifting target of selection. Most early appendages were used for feeding (e.g. jellyfish), sensory perception (e.g. the tentacles of early marine snails) or movement (e.g. trilobites). It was only later that some appendages became manipulative, notably cephalopod arms-with-suckers and primate arms-with-hands. The selection for manipulative appendages was probably different in the two cases: ways of feeding in the case of cephalopods and ways of moving in an arboreal habitat in the case of primates. Neither of these involved selection acting on intelligence *per se*.

What about features that are especially evident in the apes, and particularly the lineage that led to humans: advanced cephalization, social living, and communication via spoken language? What was selection *for* in this case? In other words, what was it about these features that improved survival in the lineage that led to *Homo sapiens*? Perhaps the best way to think about this is to consider the difficulties faced by an unarmed ape, recently descended from the trees, trying to make its way on the surface of the land, where both potential prey and predators ran faster than it did. The ability to make stone tools and to light fires would undoubtedly have aided survival. Such attributes went hand in hand with a large brain and intelligence. Perhaps here intelligence itself was finally the target of selection.

Let's now turn from Earth to other inhabited planets – in particular to those with animals. Again, as with plants, on any one planet it might be a case of *one or more* such kingdoms. Given the existence of autotrophs – organisms that use light or inorganic chemicals as their source of energy – there will always be a role for heterotrophs – organisms that eat the autotrophs (and/or each other). Evolution of heterotrophs will often be associated with evolution of mobility. This isn't inevitable, as evidenced by Earth's fungi, but it's highly probable. Evolution of powered movement is likely to be associated with the origin of contractile tissues (muscles) and a system of coordinating these (nerves).

As soon as there are multicellular mobile heterotrophs – animals – there will be selection for improved mobility. Given the advantages in this respect of a bilaterally symmetrical body form, such forms should be expected to arise often – in other words on many if not most inhabited planets – at least given sufficient time. Once they've arisen, selection for paired appendages and cephalization will follow. Basic intelligence will evolve, due largely to selection for more flexible patterns of movement. In some lineages, intelligence will reach the levels seen in cephalopods, birds, and mammals.

What then? What about extraterrestrial equivalents of the great apes, including humans? What about tool-using, social-living, language-communicating beings on other worlds? This might be a level of detail too far, in terms of thinking about the extent of parallel evolution on other planets, for at least two reasons. First, as I mentioned earlier, a planet without trees is unlikely to have apes. Perhaps, on such a planet, a lineage of octopus-like creatures will invade the land, just as their snail cousins did on Earth, and evolve human-level intelligence and a technological civilization (which their snail cousins could not). Second, any planet that's significantly younger than Earth might not have had time to evolve ape-like animals. The apes of Earth arose a mere 20 million years ago. If the history of life on Earth is four billion years, the history of apes fits into the last half of one percent. On planets whose evolution proceeds at a broadly similar pace to that on Earth, there will be no apes, and therefore no humans, after the first one, two, or three billion years.

These issues aside, on a middle-aged planet with forests, would we expect to find an ape-like branch of its evolutionary tree, running broadly in parallel to that on Earth? The best answer to this is 'maybe'. Humans are not inevitable, as argued by Conway Morris. But neither are they impossible on planets other than Earth. In the end it all comes down to probabilities. If there really are about a quintillion inhabited planets in the observable universe, as I argued in Chapter 4, then the likelihood is that there will be humanoid forms on some of them.

From the practical perspective of making contact with intelligent forms elsewhere, it's probably true to say that the nature of their technology is more important than the nature of the life forms themselves. We have no hope of communicating with humanoids on 'exoplanet X' that are at the stone-tool stage right now, like the humans of Earth three million years ago. In contrast, we might have a reasonable chance of communicating with life forms on 'exoplanet Y' who look very unlike us but who share our ability to send and receive radio signals. So, in the final chapter, technology will come to the fore.

8 Intelligence – Here and Elsewhere

Stages in the Life of a Planet

With the exception of planets orbiting the most massive and luminous stars, planetary lifespans are measured in billions of years. Evolution on Earth has taken about four billion years so far, and probably has about another two or three billion to run, depending on when our ever-brightening Sun eventually boils away all our surface water. In the absence of evidence to the contrary, it's probably a good idea to assume that evolution elsewhere takes billions of years too. It's hard to imagine an evolutionary process in which intelligence is an early result rather than a late one. So, to look for intelligent alien life, we need to concentrate on planets that aren't too young. Earlier, I suggested that good yardsticks for planetary age when looking for photosynthetic or intelligent life were about two and four billion years, respectively. In general, we can imagine at least four stages in the life of a planet – no life at all, chemosynthetic life only, a stage characterized by a mixture of forms of energy acquisition including photosynthetic life, and a final stage that also includes intelligent life. In the present chapter, we're concerned with the final one.

Of course, this scheme is a simplification. In particular, it omits the latest stages in the history of a planet – the ones we haven't yet reached here on planet Earth. There must be at least one further stage after intelligent life. Perhaps the later stages of the sequence will be a mirror image of their earlier counterparts – this would happen if intelligent life dies out first, and then all photosynthetic and heterotrophic life, perhaps leaving some extremophile chemosynthesizers to be the last to go extinct. Alternatively, if all forms of life go extinct together, there will simply be one final stage – 'No Life II', it might be called. But the exact nature of the demise of life on inhabited planets can be ignored. In the search for intelligent life, it makes sense to avoid planets that are close to dying, just as we avoid those that have recently been born.

It's clear, then, what *stage* of planet to focus on, in order to look for intelligent life. And we know what *kind* of planet to focus on – rocky ones orbiting in the habitable zone of their host stars. But what exactly are we looking *for*? What would we take to constitute evidence of intelligent life with a technological civilization? The short answer, of course, is 'technosignatures'. But, as with 'biosignatures', this is a slippery term. Some authors take a broad view, and include anything that might be a sign of technology. Others take more restrictive views, for example including industrial modification of atmospheres but excluding radio signals. I don't see any good reason to make exclusions, so the usage I adopt here is the broad one.

The search for intelligent life thus becomes the search for advanced technology, together with its intentional products and its unintentional side effects. Naturally, this is a pragmatic choice, not an ideal one. We would love to know if there are planets 'out there' where intelligent apes are making and using stone tools. Equally, we would love to know if there are planets with technologically minimal civilizations whose citizens indulge in pursuits such as mathematics, natural history, and philosophy – like Earth's ancient Greeks. But realistically we have no way to approach such issues beyond the solar system, and by now it's clear that our system has no such forms of life anywhere other than on Earth.

There's a hidden danger in our pragmatic choice of how to search for technology that's a product of the activities of intelligent life forms. Perhaps, just as there may be life that is 'not as we know it', there may be alien technology to which the same label can be applied. But, just as with 'contrasting life', the problem of 'contrasting technology' is that we don't know if it's possible. In the same way as it's not clear how life could operate on the basis of silicon rather than carbon, it's hard to see how technology could be produced without metals. And it's hard to imagine how alien beings would try to communicate their existence to the universe at large – assuming they wanted to – other than by means that we know of, such as radio signals or lasers.

So here's the strategy of this chapter. Having examined the sort of planets to focus on, and the sort of evidence we're searching for, we'll have a look at the evolution of the human brain and the evolution (in a broader sense) of the technology that it has enabled on Earth. Then we'll consider possible parallels elsewhere. After that, we'll focus on the search itself, and related issues such as the Fermi paradox. Here goes.

Human Evolution on Earth

We've already examined the early bases of the evolution of intelligence in the animal kingdom, including movement, nervous systems, bilateral symmetry, manipulative appendages, and cephalization. The account that follows begins where those previous accounts finished – with a mobile bilaterally symmetric animal that's highly cephalized and has appendages that can be used to manipulate various objects. In particular, it begins with chimpanzees.

The degree of similarity between the genomes of humans and chimps is about 99%. Without a doubt, chimps are our closest living relatives. While present-day humans all belong to a single species, *Homo sapiens*, present-day chimps belong to two – the 'common chimps' and the 'pygmy chimps' or bonobos. These two chimp species diverged from each other very recently – between one and two million years ago. The divergence of human and chimp lineages happened earlier than that. This key lineage fork took place sometime between five and ten million years ago. It happened in Africa, though exactly where in that vast continent is still a matter of debate. So we know both the when and the where of the fork roughly, but not exactly.

The lineages which led from that fork to modern humans branched many times, but the resulting species are all extinct, except for our own. With regard to the evolution of human intelligence, the main usefulness of fossils belonging to the other species is to provide us with data from which we can attempt to reconstruct the tempo of evolution of this key human feature. However, many of the fossils are so fragmentary – a broken bit of jawbone and/or a few teeth – that they tell us little or nothing in this respect. The ones that help are those that include enough bits of braincase to allow an estimate to be made of cranial capacity (brain size), which can be thought of as a measure, albeit a very blunt one, of the level of intelligence of the animals concerned.

Brain sizes in primates are measured in the same way as car engines of the pre-electric era – in cubic centimetres (abbreviated to either cm^3 or cc). The Fiat Cinquecento is a good marker of small engine size – 500 cc, as is implicit in its name. A marker of a reasonably large engine size (by European but perhaps not American standards) is a Jaguar XE, with 2000 cc, alias 2 litres. What's measured by such figures is the combined operational volume of however many cylinders there are in the engine concerned. Modern chimps have a brain size of about 400 cc when adult – a bit smaller than the engine of the little Fiat. In contrast, modern adult humans have a brain size of about 1400 cc – between three and four times larger. Assuming that the last common ancestor of human and chimp had a brain size not too different from modern chimps, which is probably about right, the evolution of our lineage has resulted in the addition of a whole litre in brain size over a few million years. That's impressive.

What do we know about the pattern of this increase? It seems that most of it occurred in the last two million years (Figure 8.1), with the proliferation of species within the genus *Homo*, of which there are thought to have been about a dozen. I won't name most of these, to avoid drowning you in unnecessary jargon. Here's a simplified version of an extremely complex story. Between three and two million years ago, the tool-making *Homo habilis* (handy man) appeared, with a brain size of about 600–700 cc. Between 2 and 0.5 million years ago, *H. erectus* (upright man) had a brain averaging about 900 cc. The litre mark was passed about 0.5 million years ago with *H. heidelbergensis* (whose name reflects where fossils of it were first found). And brains averaging more than 1250 cc appeared with the evolution of our own species, and the Neanderthals, who became extinct less than 50,000 years ago.

However, the evolution of brain size is far from being a monotonic trend going inexorably upwards. The three most recent species of *Homo* are us, the Neanderthals (whose species status is controversial), and *H. floresiensis*, otherwise known as the hobbit – a species endemic to the Indonesian island of Flores, as its name suggests. Individuals of this species had brains of only around 400 cc, in other words very similar to chimps. Part of the explanation for their small brains is their small body size, but this isn't the whole story. If we look at information on relative brain size – that is, brain size per unit body size – the hobbits were more similar to *H. erectus* than to *H. sapiens*.

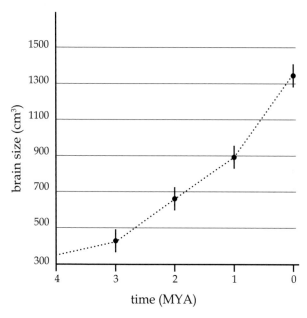

Figure 8.1   The approximate pattern of increase in human brain size over the last four million years, from a starting point of about 400 cm³ to today's value of about 1400 cm³. Of course, evolution is a messy process and there are various complications to the suspiciously neat curve shown, of which two in particular are worthy of mention here. First, a temporal series of values for average brain size is just that. How it connects up between the actual data points, and in particular whether it connects up in a gradual or punctuational manner, cannot be specified until there are data points from all time horizons. Second, the pattern shown applies to a particular lineage – the one leading to *Homo sapiens*. Other lineages, in particular that which led to the recently extinct 'hobbit' (*Homo floresiensis*), involved decreases, and also remind us to consider the ratio of brain size to body size. MYA, millions of years ago.

Another complication is that brain size varies from one individual adult to another. The above figures provide a rough idea of the *average* adult value for each species, and should not be interpreted as constants. At every stage in the evolution of a group of animals, all measurable characters exhibit variation, and brain size in species of *Homo* is no exception. There are other complications too, including interbreeding between 'species', thus rendering the use of that term dubious. The best-known case of this involves modern humans and Neanderthals (including the Denisovan lineage), only a few tens of thousands of years ago.

What drove the amazing evolutionary increase in brain size that added a litre to our cranial capacity, via modification of brain development? Many factors rather than one, to be sure, with three being especially important, as follows. First, return from the trees to the ground, with subsequent freeing up of the hands for a manipulative role, and the development of this role through the use of tools. Second, the evolution of language, and dealing with the complexities of communication that this allowed. Third, living a social existence. Although this was already common in African apes by the time humans first arrived on the scene (it's less so in the Asian orangs), the three-way combination of societal living, use of tools, and elaboration of language was a powerful driver of selection for bigger brains.

However, a caveat is necessary here. The above hypothesis of the main factors involved in driving the amazing increase in brain size in the lineage that led to *H. sapiens* is a plausible one. But it is, nevertheless, only a hypothesis. Such hypotheses about the *reasons* for evolutionary changes, which can be called 'adaptive' or 'selective' in the sense of relating to adaptation or natural selection, are difficult – some would say impossible – to test. This is true regardless of the character concerned, be it brain size, body size, or something else. It's clearly the case that present-day horses are significantly larger than their ancestors of 50 million years ago, which were similar in size to foxes. But the reasons why they got larger, in other words the environmental challenges to which their size was adapting, are less certain. Likewise, while modern human brains are significantly larger than those of our ancestors of five million years ago, the exact reasons why they are so much larger are open to debate.

The difficulty of testing adaptive hypotheses has led some critics to write off this whole approach as speculative 'adaptive story-telling'. They argue that evolutionary *patterns* can be 'read from the rocks' whereas the selective *processes* that generate them cannot (refer back to Chapter 7 for the distinction between evolutionary pattern and process). I agree with such critics in their identification of the problem of testability, but not with their overall negativity about adaptive hypotheses. Biologists are interested not just in *what* happens in evolution, but also in *why* it happens. The difficulties inherent in testing adaptive hypotheses should lead us to be cautious when adopting this approach, but not to abandon it, and with it the search for answers to those all-important 'why' questions.

Human Technology

The earliest known stone tools were manufactured in the period between 3.5 and 3.0 million years ago – long before the evolution of *Homo sapiens*, indeed even before the evolution of the earliest members of the genus *Homo*. Stone, together with wood and bone, continued to be the basis of our ancestors' tools all the way from then to a mere few thousand years ago, when the so-called Bronze Age began. Technology evolved and improved much over this period of more than three million years, of course, but by the time the last ice age ended all tools were still based on stone, bone, or wood, rather than metals. All the evidence we have of this ancient technology is in the forms of the artefacts themselves, except in the very latest stages, where some were depicted in cave paintings – though they feature rarely compared to paintings of animals.

It's a strange fact, but the last ice age finished only about 10,000–12,000 years ago, and yet all recorded human history is within post-glacial times. In evolutionary terms, this is a mere instant. We'll look at examples of the evolution of human technology over this period. Of course, the evolution of technology is a different process than the evolution of the organisms making the tools. Tools don't reproduce, so their evolution is not driven by natural selection. But it's an evolutionary process in a general sense, regardless of that fact. Let's now make a comparison of the nice round numbers of 10,000, 1000, and 100 years ago.

By 10,000 years ago stone tools still prevailed, but they were accompanied by other signs of progress in human societies. Agriculture had begun, including the cultivation of a few species of plants and the domestication of a few species of animals. But there were not yet any sizeable towns or cities. The invention of the wheel lay a few thousand years into the future, as did the first written records. Writing is thought to have begun between 4000 and 3000 BCE, or, if you prefer, between 6000 and 5000 BP (before the present).

By 1000 years ago, there was everything you can picture from recorded events of that time, such as the Battle of Hastings (1066) in the Norman conquest of England. These include towns, castles, bows and arrows, cavalry forces, swords, armour, and many other accoutrements of battle; plus their domestic counterparts, including furniture, candles, cutlery, and so on. But no sign yet of electricity, printed books, or spectacles. The first university was established at Bologna in 1088.

A hundred years ago, in the 1920s, we had entered the radio age. There were trains and steamships, road vehicles that we would recognize as cars, and powered flight – which began with the Wright brothers' first flight in 1903. There were large cities – the population of the biggest of them at the time, London, was over six million. The First World War saw the use of tanks, machine guns, and submarines. Morse code was in widespread use. There were cinemas, and work had begun on television sets. Penicillin, the first antibiotic, was discovered in the late 1920s. And many parts of the world were becoming connected up by telephone.

Over the last century, the rate of change in human science and technology has been extraordinary. We've gone from early air travel to landing space probes on asteroids and comets. From the first phone systems to the Internet. From the discovery of penicillin to the Human Genome Project. From uncertainty regarding whether 'spiral nebulae' were inside or outside the Milky Way to the discovery of thousands of galaxies.

Thinking back to how comparatively little our technology advanced between 100,000 years ago and the end of the last ice age, some 90,000 years later, those who argue against a slow-then-fast, or accelerating, pattern of change – and there are some – are hard to believe. An exponential rate of increase can be argued against because that's a very specific shape of curve, and few events in the real world fit it perfectly. But a more general claim of a slow-then-fast pattern is difficult to refute. This conclusion leads to an interesting problem. There's a mismatch between this pattern of progress in our technology and the pattern of evolution that characterized our increasing brain size. By the time technology began to speed up appreciably, around 10,000 years ago, our brain size had plateaued. Indeed, the plateau had probably begun much earlier – maybe about 100,000 years ago.

Should this mismatch be a cause for concern? Well, I'd say 'no' in one respect but 'yes' in another. A lag between evolution of the mental capacities that a large brain enables and the appearance of a technology that goes beyond stone tools, which is ultimately a product of such a brain, is to be expected. You can't go straight from hand axes to space telescopes. However, this mismatch does direct our attention back once again to the question of why such large brains evolved in the first place, given that most of the tangible benefits to which they led were far in the future. Natural selection, as we know, works only with combinations of the products of the past and the pressures of the present – it cannot look into the future, not even by a small amount of time, such as a century. Thus we return to the adaptive hypothesis that our large brains evolved for reasons that included the ability to make stone tools, the origin of language, our living in increasingly large social groups, and the connections among these three things. Is it believable that brains capable of understanding the nature of the universe and the evolutionary process that made them arose for those reasons and those only?

That's a tough question, and we can't yet really see our way towards a satisfying answer to it. Other forms of Darwinian process were probably involved, notably sexual selection. Here, bigger brains, or features associated with them, are favoured in terms of mate choice, and hence in the reproductive component of fitness rather than the survival one. But was anything more radical involved? Darwin's colleague Alfred Russel Wallace thought that natural selection couldn't be the explanation of advanced human mental abilities. He was a biologist through and through with regard to all of the animal kingdom except for a single species – our own. With regard to humans, Wallace was a spiritualist. Was he right to exclude natural selection as a cause of mathematical ability, musical ability, and so on? I think not. What about those who believe that aliens arrived and hybridized with proto-humans, hence leading to the mental abilities of modern man? Might they be right? Again, I think not. I wouldn't advocate keeping an open mind in relation to either of these two notions. But keeping an open mind about whether our best current adaptive explanations are correct is most certainly to be recommended.

Intelligence and Technology Elsewhere

Many features of life on Earth have evolved multiple times from different starting points, which gives us at least a pointer to the probability that their evolution is somehow 'easy' – even if it takes millions of years – and might be expected to characterize the evolutionary processes of other planets too. This is even true of early-stage intelligence, since that evolved more or less independently in invertebrates (cephalopod molluscs) and vertebrates (birds and mammals). But it's not true of human-level intelligence and the technology that goes with it. My suggestions below are thus very speculative – much more so, I think, than suggestions about 'the basics of life' on other planets. Given this fact, it's best that I keep the current section brief.

What can we be most certain – or least uncertain – about in relation to human-level intelligence (or beyond) associated with an advanced technology on other worlds? In cases where extraterrestrial life is parallel to Earthly life in its basics – i.e. carbon-based and constructed of cells – I'd say that intelligent life forms on the planets concerned are likely to be large, multicellular, and mobile. They're likely to have nervous systems that have become highly cephalized, in other words to have large brains. They probably have manipulative appendages of some sort, probably in pairs, though not necessarily just one such pair. They probably live on land rather than in water. They probably live socially and have some form of language that allows complex communication between individuals.

I'm aware that there are alternative views. This is hardly surprising, given that it's a huge subject and as yet we lack any information from other inhabited planets. Some people like to think in terms of 'plant intelligence' on Earth, though I think this is stretching the definition too far. But there may be planets on which the clean separation between animals and plants that occurs here doesn't apply. John Wyndham's large mobile predatory plants – the triffids – are only fiction here on Earth, but they may be a reality somewhere else. If they exist, they may have tissue-types that don't exist here, including some basis for intelligence that doesn't equate with our animal-based idea of a nervous system. That said, however, I still consider intelligent alien 'animaloids' to be more likely than plant-based counterparts.

But will intelligent aliens be humanoid in form? That's a harder question. I think the best way to deal with it is to think very widely in terms of inhabited planets. Our scope should encompass the whole of the observable universe rather than just our local part of it – the Orion spur of the Milky Way. If there are many intelligent creatures in the universe, scattered across countless inhabited planets, then out of all the planets on which evolution has produced high intelligence, there are probably some that have humanoids and some that have very different intelligent forms. I mentioned earlier the possibility of a planet on which an octopoid life form emerged from the oceans and colonized the land. It would have to solve all manner of adaptive problems to be successful in such colonization, not least of which would be to acquire oxygen from air rather than water. But since their gastropod cousins, the land-snails, managed to evolve lungs from gills, why should this not be possible in cephalopods too?

Given the desirability of keeping an open mind, it's also worth considering another possibility – that multicellular animaloids on some planets might not overlap at all with animals on Earth in terms of their morphology. The trouble is that to envisage such a non-overlapping fauna we have to invent seemingly improbable forms. For example, perhaps there is an animal kingdom somewhere all of whose member species, including those with intelligence, have body forms characterized by hexagonal symmetry. It doesn't seem likely, given that bilateral symmetry appears to be a prerequisite for high intelligence here on Earth. However, given so many possible inhabited planets, we can't rule out such possibilities altogether.

In the end, though, the body forms of intelligent beings on other planets might not matter in terms of our detecting them, any more than the body forms of photosynthesizers matter in this respect. Technosignatures might be similar, despite being sent by beings that look nothing like us, just as oxygen biosignatures might be similar, regardless of whether they're created by tiny bacteria or lofty trees. However, this view requires there to be constraints on technology that in a sense transcend the body forms of the beings who produce them. That issue takes us from the evolution of intelligence to the evolution of technology itself.

Let's focus on detecting radio signals that have been deliberately sent out into space by intelligent beings on other planets. Might we be trying to detect the wrong type of signal, having been overly influenced by the messages we send out ourselves? In conversations about the possible nature of signals sent out by aliens, I've often heard the following suggestion. Perhaps intelligent aliens on some planets are much more advanced than we are in terms of their science and technology. If this is the case – and it's very likely to be – then perhaps they'll have discovered some way of signalling that's somehow 'better' than radio, or indeed than the electromagnetic spectrum in general. They may be broadcasting such signals in our direction, but we can't detect them because our science is too primitive.

Again it's a case of something that seems unlikely, but can't be ruled out. We still don't understand the nature of dark energy. And we weren't able to detect gravitational waves before the twenty-first century, even though Albert Einstein had predicted them about a century earlier. There may be other forms of energy of which, as yet, we don't even have an inkling. While this possibility can't be dismissed, it doesn't help our search for alien intelligence in practical terms. As long as there are some intelligent life forms somewhere that are using radio signals to broadcast their existence, that gives us something to search for using our current technology. Let's now examine the search so far.

From Project Ozma to Breakthrough Listen

The search for radio signals from technological civilizations is a major part of the overall SETI endeavour. It started in 1960 and continues today. I mentioned the first SETI undertaking – Project Ozma – in Chapter 1. Many recent and current SETI searches are conducted under the banner of Breakthrough Listen – one of the Breakthrough Initiatives, two others of which we've already met (Breakthrough Enceladus and Breakthrough Starshot). The methods of searching have evolved over the sixty or so years since Ozma, but the main difference between early and recent searches is not so much in their methods as in their scale.

In Project Ozma, Frank Drake and his collaborators focused their search on two stars – Tau Ceti (which we met in Chapter 6) and Epsilon Eridani – over a period of four months. There was a bigger-scale follow-up project (Ozma II) which involved focusing on 670 stars over four years in the early 1970s.

But this larger sample of stars is dwarfed by the scope of Breakthrough Listen, which includes about a million stars in the Milky Way, as well as some other galaxies. This broad-scale search began in 2016, the year after the official launch of the Breakthrough Initiatives, and is planned to last for a decade.

The scope of Breakthrough Listen is also greater than that of Project Ozma in a different way – the span of wavelengths searched. In Ozma (both I and II), the search was concentrated on wavelengths around 21 centimetres, which corresponds to a frequency of 1420 megahertz (MHz). Such a focus had been suggested in a paper by Giuseppe Cocconi and Philip Morrison in 1959. Their rationale was that intelligent beings would probably know that there is a natural source of emission at this wavelength – hydrogen atoms. They argued that an alien civilization wishing to broadcast its existence to other civilizations – including our own – might well use this particular frequency/wavelength to maximize the chances of finding someone 'listening'. While this is a reasonable argument, the broader the range of frequencies examined for possible signals, the greater the chance of detecting one. So Breakthrough Listen searches over a band of wavelengths that is several times that of Ozma. It also includes a search for signals sent by laser rather than radio.

Many SETI searches have been carried out in the period between Project Ozma and Breakthrough Listen. They have been conducted by various investigators based at a range of institutions, including the SETI Institute and the Berkeley SETI Research Center, both based in California, the latter being host to Breakthrough Listen. What have they collectively discovered? Well, there have been a few promising possibilities, but none has stood up to scrutiny – so far. At the time of writing, the most recent of these was BLC-1 (Breakthrough Listen candidate 1), which was discovered in 2020 by analysis of data from the previous year. This radio transmission seemed to be coming from the direction of the Alpha Centauri system. However, the most likely explanation now seems to be human radio signals being mistaken for those coming from space.

Over the same period as we've been searching for incoming radio signals, we've been sending out some of our own. So far, we've sent between ten and twenty – the exact number depends on how you define a single message, since some have been broadcast more than once. The first – the Morse message – was sent in 1963, the most famous – the Arecibo message – in 1974. These were both unusual in that their targets were respectively very close (Venus) and very distant (a star cluster more than 20,000 light years away). Most that have been sent since then have been directed towards stars with known or suspected planets. Their distances range from about 40 to about 400 light years, so no reply could be expected any time soon. There's one exception – a message sent towards Luyten b, a planet in the habitable zone of Luyten's Star, which is a mere 12 light years distant. This message – Sonar Calling – will arrive in 2030, so a reply could arrive here by 2042, thus perhaps confirming Douglas Adams' wonderful idea that 42 might be the meaning of 'life, the universe, and everything'. However, since the star orbited is a red dwarf, Luyten b is probably tidally locked to it, so excitement about a reply may be unwarranted.

The negative SETI results to date should not be a cause for pessimism in relation to detecting signals from alien civilizations. Our search for extraterrestrial intelligence is still in its infancy. And the critical assessment of candidate signals, in order to consider all possible kinds of source, is to be applauded. It's very important that the cautionary message about false positives in the paper by James Green and his colleagues, which I mentioned right at the start of the book, is taken seriously. This is even more true of possible SETI results than of possible findings of extraterrestrial microbes, as in the recent study of the Venusian atmosphere that we examined in Chapter 5. If the possibility of alien bacteria can be the cause of media hype, think how much greater the hype could be in the case of a claimed discovery of radio signals from an alien civilization.

So the current situation is that the search goes on, and indeed since the start of Breakthrough Listen it goes on at a much greater rate than before, due to the huge amount of funding involved – about $100 million. There are still some years of this project left to run, and there are likely to be follow-up projects too. I'm optimistic that one day we'll receive a signal that isn't an artefact, one that does actually tell us that we are not alone as intelligent beings in the universe. However, it's hard to put a likely timescale on this momentous discovery. Patience is the name of the game.

Other Technosignatures

The radio and laser messages from alien civilizations for which SETI usually searches are deliberate attempts to make contact. However, there may be many civilizations that decide not to broadcast their existence into space, perhaps deeming it to be a risky venture. But such civilizations may be detectable by other (unintentional) technosignatures. These signatures could be of several different types, including: radio 'leakage' (as opposed to deliberate signals); industrial gases, such as the damaging CFCs in Earth's atmosphere, which are now thankfully on the decline; and metal-based artefacts. In this last category are both large-scale astro-engineering projects such as Dyson spheres (see below) and spacecraft sent out beyond the planetary system of the civilization concerned.

We noted earlier that NASA's two *Voyager* spacecraft are now leaving the solar system, and they're not alone. *Pioneer 10* and *11*, also launched in the 1970s, are doing likewise. There's a difference though. The *Voyager* craft are still able to transmit radio signals to us, while contact with the *Pioneer* craft has been lost. This is relevant to us, but not so much to potential alien discoverers of these craft in the distant future. If any of them are captured by the gravitational pull of a star, and detected by a civilization on one of the planets orbiting it, their structure will immediately reveal them to be artefacts created by intelligent beings, rather than natural objects. As we saw earlier, one hypothesis of the nature of 'Oumuamua is that it represents space technology produced by an alien civilization – an equivalent for us of an alien *Voyager* (or a discarded part thereof) that has made an interstellar journey from their system to ours. This is not the consensus view among scientists, but it's an interesting idea.

A Dyson sphere is a hypothetical type of megastructure that consists of a series of interconnected panels surrounding a star. The idea is that a technological civilization on one of the star's orbiting planets might build such a structure to increase the amount of energy it can acquire from the star concerned. It's named after the British-American physicist Freeman Dyson, who proposed it – though such concepts had previously featured in some science fiction writing. There are lots of variants on the idea – for example a Dyson swarm of smaller quasi-independent modules surrounding the star.

While such a construct is conceivable, its scale does raise questions about its feasibility, even to a civilization that's much more advanced than our own. The sheer amount of material involved might be prohibitive. Perhaps more likely than a stellar-scale megastructure would be a planet-scale one. An example would be a partial planetary shield used by advanced civilizations to reduce the amount of incident sunlight when their local star becomes too luminous for the continuation of life – a 'boiling the oceans' scenario, which is an all-too-real prospect for Earth and other planets that start off in habitable zones. But even if these exist, they might be difficult to detect from a distance of many light years.

In my view, none of these other technosignatures are as promising targets for SETI research as radio (and laser) signals. I suspect that the first evidence we obtain of intelligent extraterrestrial life will be an incoming radio message, perhaps one sent speculatively rather than a reply to one of our outgoing signals. But, given the uncertainty about when we might receive such a message, there's a good chance that, by the time we do, we'll already have evidence for the existence of photosynthetic extraterrestrial life via a strong oxygen signal from the atmosphere of an exoplanet in the habitable zone. This raises the interesting possibility that the two kinds of searches could interact. Planets with probable photosynthetic life would make good targets for future outgoing radio messages. This is because such planets would have a higher probability of having intelligent life than planets from whose atmospheres we detect no oxygen signals. One critical point in such a case would be to have a reasonable estimate of the age of the system concerned. On Earth, the gap between the Great Oxygenation Event and the advent of human technological civilization was about two billion years.

The Fermi Paradox

Everyone knows what the Fermi paradox is – or do they? Roughly stated, it's the apparent contradiction between the statistical probability that there are many technological civilizations in the observable universe and the fact that (so far) we've not heard from any of them. I will argue that in fact there may be no contradiction at all. But before I do that, let's flesh out the background to the claimed paradox.

Enrico Fermi was an Italian-American physicist who was born in Rome in 1901, was educated in Pisa, and was appointed a professor at Rome's Sapienza University at the age of 24. He was awarded the Nobel Prize in Physics in 1938, for work on radioactivity. Having gone to Stockholm with his wife Laura to collect the prize, Fermi did not return to his native Italy, but rather went to New York, took a job at Columbia University, and applied for residency. The reason for his transatlantic move was the introduction of the 'race laws' (*le leggi razziali*) by the Italian fascist government. These laws, the first of which came into force in 1938, were a threat to Laura, who was Jewish.

Fermi was involved in many research endeavours in the USA, including the Manhattan Project – the R&D undertaking that led to the production of the first atomic bombs. During the project, he was based at Los Alamos National Laboratory in New Mexico. After the war, he became a professor of physics at the University of Chicago, but he continued to spend time at Los Alamos. It was there, in the summer of 1950, that a lunchtime conversation between Fermi and some colleagues led to the recognition of what we now call the Fermi paradox.

The story goes something like this. The scientists were chatting about something (it's not clear exactly what) when Fermi suddenly posed a question, along the lines of 'Wait a moment, where is everybody?' Apparently, his colleagues all immediately knew that he was talking about aliens – intelligent ones. Various *calculations* suggest that the universe should contain many planets with intelligent life. Yet there is no believable *evidence* that Earth has ever been visited by aliens, or that Earth scientists have ever received messages from alien civilizations via radio signals or any other means of communication.

This juxtaposition of 'calculations' suggesting that intelligent aliens should be common in the universe and 'lack of evidence' for the existence of any such life forms is problematic as the basis of a paradox, because neither of these things is terribly impressive. The 'calculations' of the number of planets with intelligent civilizations involve lots of assumptions. The 'lack of evidence' is really just a statement that our own civilization, which is less than a century into the space age, has so far not had any contact with aliens. This is perhaps unsurprising, especially if alien civilizations are widely spaced, as seems likely.

Let's approach the Fermi paradox from the starting point of the more pessimistic of our two estimates (from Chapter 4) for the number of inhabited planets in the Milky Way: ten million. Recall that this estimate applies to the number of planets with *any* type of life, and in many cases it will be entirely microbial. We now need to reduce our starting figure appropriately to a lower figure representing the number of planets with *intelligent* life. But what is the magnitude of 'appropriately'? How do we even begin to quantify such a reduction? As I said above, these calculations involve many assumptions. We'll now do the calculations, noting the assumptions along the way.

I'm going to take a two-stage approach to the reduction in the number of inhabited planets: the first step is going from any kind of life to animal life, the second from animals to intelligence. In adopting this approach, I've already made one assumption – that intelligence can only arise in animals or animal-like forms, and not, for example, in bacteria. Here's a guestimate of the fraction of inhabited planets that have animals: 10%. But where does this figure come from? My rationale goes like this. On Earth, microbes have been around for about four billion years, animals for only about 0.6 billion (or 600 million, if you prefer). Thus 'animal time so far' is 15% of 'microbial time so far'. But any such calculations operate in an order-of-magnitude manner at best, as we saw in Chapter 4. Also, as time moves on, the fraction will gradually increase, assuming that both microbes and animals survive. However, towards the end of Earth's (or any planet's) period of existence, conditions become harsh, primarily because of the boiling off of surface water by the ever-brightening host star. Since microbes seem to generate extremophile forms more readily than animals do, animals may die out first, so the fraction of the planet's inhabited history for which they've been around may begin to fall again. The inhabited planets that now exist in the Milky Way are at a variety of stages in their evolutionary histories. So what we need is an order-of-magnitude guestimate of the average fraction of those histories that animals have been around for, and 10% seems a good choice. It's certainly better than 1% or 100%. But whether it's better than 5% or 15% we really have no idea.

Now for the second stage of our reduction in the number of inhabited planets – from those with animals to those with intelligence. This calculation is arguably even harder than the previous one. A possible approach would be to say that since animals have been on Earth for about 600 million years, and the human lineage 'began' (split off from the chimp one) about six million years ago, we should use a figure of 1%. However, recall (from Chapter 1) our pragmatic use of radio capability as a definition of intelligent civilizations. Humanity has only been radio-capable for about 120 years. If we use that number instead of six million for the start of 'intelligence', our guestimated proportion of 1% drops by several orders of magnitude, and becomes vanishingly low. But the same temporal principle that applied in the first stage of our calculations applies here too: the proportion will slowly rise as time goes on, assuming that both intelligent and unintelligent animals survive together. So what's a good guestimate for the overall proportion of planets with animal life that also have intelligent life characterized by technological civilizations? It's likely to be less than 1%, but by how much? One in a thousand (0.1%) still seems overly optimistic, so let's go with one in ten thousand (0.01%).

From a starting point of ten million planets with life, we now have guestimates of a million with animal life and a hundred with intelligent life. There are huge errors involved in generating both these figures. However, let's accept them and imagine how our hundred civilizations might be distributed in space.

For this purpose, let's simplify the Milky Way as follows, starting with the fact that its diameter is about 100,000 light years. We'll pretend that instead of being roughly circular (with a central bulge and spiral arms), it's actually square. And we'll pretend that instead of having some finite depth it's effectively flat. We'll assume that stars are distributed evenly across the square, each of whose sides measures 100,000 light years. Now we'll divide the square up into 10 rows and 10 columns, giving 100 vast cells of space, each of which is 10,000 × 10,000 light years (Figure 8.2). Let's imagine that the 100 stars hosting planets with intelligent life are regularly spaced (unlikely!), with each being found at the centre of one of these giant cells. Our own solar system is located at one of these central points. Assuming that we're not at an edge location in the overall hypothetical square, which makes sense because we're not at a peripheral location in the overall circle/spiral of the real Milky Way, then we have four nearest civilizations, as shown in Figure 8.2. Each of these is 10,000 light years away.

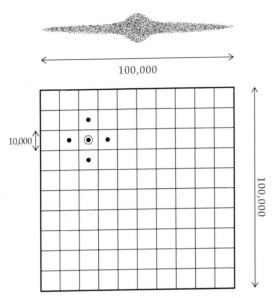

Figure 8.2 One possible pattern of distribution of intelligent life across the Milky Way, shown in simplified grid form (see text). Assuming that intelligent life occurs on only a small subset of inhabited planets generally, there will be hundreds or thousands of light years separating one technological civilization from another. The pattern shown involves an inter-civilization distance of 10,000 light years. If something like this distance applies in reality, then there are implications for the likelihood of humans discovering other civilizations.

We can now think of the 'lack of evidence' part of the Fermi paradox in the context of a spatial model that may not be so very far from reality. Are we likely to have received detectable radio signals or spacecraft visits from a civilization that's 10,000 light years away? I'd say that the answer is 'no'. One reason for this conclusion is the large number of stars and planets between us and our nearest intelligent neighbours. There may well be more than a billion such stars, the vast majority of which will have multiple planets. That should keep an intelligent civilization busy for a very long time, in terms of exploration of their local area of the galaxy, before they discover the Earth. Perhaps the 'Fermi paradox' isn't paradoxical at all.

Naturally, it's easy to construct a counter-view in which Fermi's paradox is very real. If I had started the above calculations with the more optimistic of Chapter 4's two estimates of the number of inhabited planets in the Milky Way, Figure 8.2 would have looked very different. Another possibility is that contact with alien civilizations is rendered far more probable than I have imagined because some of them may have a distinctly expansionist outlook and a penchant for robotic self-replicating spacecraft or something broadly equivalent to that. Such craft could effectively extend the reach of the civilization concerned. Also, advanced civilizations may discover ways to escape Einstein's universal speed limit or discover wormholes that are short-cuts from one part of the galaxy to another.

Implicit in the above estimation of the possible distribution of technological civilizations in space is their possible duration in time. The longer civilizations last, the more of them there will be at a given moment – for example the present – other things being equal. Hence the closer together they will be. Frank Drake was well aware of the importance of the duration of intelligent life when he came up with his famous 1961 equation to estimate the number of civilizations in the Milky Way, the story of which is told in the book *Is Anyone Out There?* (1992, co-authored by Drake and Dava Sobel). In his approach, the average such duration was made explicit as a particular parameter. In my calculations above, it was implicit instead – in the fraction of planets with animals that also have intelligence.

How long will humans endure from the start of the radio age, around 1900? We've managed only a century or so thus far, and sometimes it seems like we won't get much further. In recent years, the spectres of runaway climate change and viral pandemics have come into in sharp focus. So has the spectre of nuclear war, with Russia's barbaric invasion of Ukraine in 2022. But perhaps if we survive another century things will get easier – especially if we learn from our mistakes before they overwhelm us. If we last for the next hundred years, why not a thousand or a million?

Our uncertainty about duration is multiplied enormously when we consider how long other civilizations might last – those of which as yet we know nothing at all. If the durations of individual civilizations are unknown to us, calculating their average is necessarily an impossible task. Perhaps some civilizations even last for *billions* of years. However, such a duration would take us into unknown territory, because the species upon whose technology the civilization is based would repeatedly evolve into descendant species over this vast period of time. If we humans on Earth last for such an extended period, we won't be *Homo sapiens* at the end of it.

So, given all the uncertainties involved at the current stage of human knowledge about life in the universe, the solution to the Fermi paradox remains a mystery. There have been countless suggestions, ranging from the idea that there are no technological civilizations anywhere apart from our own to the bizarre notion that intelligent aliens are already here on Earth undetected. But the solution I'm most drawn to is the one that was illustrated in Figure 8.2: rarity of planets with intelligent life as a fraction of inhabited planets in general, with correspondingly large distances separating one technological civilization from another. Where is everybody? Very far away.

Concluding Remarks

I started the book by considering the possibility that our generation may be the one to discover the first persuasive evidence of extraterrestrial life. Suppose this turns out to be true, and humanity's first such evidence arrives in a decade or so. What would the possible impacts be? We can deal with them under four headings: scientific, social, religious, and philosophical, with all of these being defined very broadly.

Scientifically, our first discovery of alien life will rank as one of the most important discoveries ever about the nature of the universe, regardless of whether it's photosynthetic or intelligent life. If it's the latter, then other possible scientific impacts abound, including those with practical import-ance. For example, information from our 'intelligent neighbours' might allow us to accelerate our research programme into the harnessing of nuclear fusion, giving the prospect of providing almost limitless clean energy for Earth. This could solve most of our problems of human-induced climate change at one fell swoop, by rendering both fossil fuels and fission fuels redundant. There might be a concern that learning such things from others is somehow cheating. But any such quibbles would surely be outweighed by the gains. Also, we might receive information from an intelligent neighbour on one exoplanet that's informative about life on others; this would apply if the civilization that sends us signals has already discovered life on other planets in its own cosmic neighbour-hood. In this way, the search for alien life might incorporate a sort of positive-feedback effect.

Socially, the discovery of photosynthetic life would cause worldwide head-lines, and would perhaps remain in the news for longer than most other items.

But it might soon be forgotten by the general public, just as the discovery of the first Earth-like exoplanet in a habitable zone has been. Not many people in the current decade can name the year in which this discovery was made – 2014, as we saw in Chapter 6 – despite the fact that they have no difficulty recalling the year, a century earlier, when the First World War began. The discovery of *intelligent* life would remain much longer in the public mind, though even that might eventually recede if our signals are intermittent and replies take 100 years or so to be received. The public panic resulting from 'first contact' that's often depicted in the realm of sci-fi is probably overstated. There is no need for governments to hide incoming signals from our sight.

The religious angle on this issue is, I think, particularly interesting. Having given many talks on extraterrestrial life at public venues, I've noticed a distinct reluctance on the part of many people who are conventionally religious to accept the possibility of intelligent aliens – though such folk seem to have few concerns about microbial alien life. The reason why their concerns take this particular pattern seems to be a feeling that there's only one intelligent life form anywhere that the presumed deity is looking after – us. As someone who ditched his native Presbyterianism as a teenager on the basis of reason, I can't really connect with this point of view. Anyhow, when we do eventually discover alien intelligence, no doubt religions will morph in some way to take it on board, without getting rid of their core beliefs, just as they have other uncomfortable (for them) discoveries, like the ancientness of the Earth, Darwinian evolution, and the electrochemical nature of our thoughts. That's the trouble with unfounded beliefs – they can't be got rid of as easily as incorrect hypotheses.

Philosophically, the discovery of alien life would have a major impact – a very positive one, at least for folk who aren't conventionally religious. I'd say that this is true regardless of whether it turns out to be photosynthetic, intelligent, or both – the last being a possibility, despite the fact that I haven't set much store by the idea of intelligent plants. Throughout human history, we've looked out into the darkness of the night sky and seen no signs of life. Since the start of the space age, we've even been able to observe lifeless extraterrestrial landscapes – notably on the Moon, on Mars, and on Titan. Despite strong arguments for alien life based on probabilities, we feel alone as mortal beings surrounded by a vast expanse of lifeless space, apparently punctuated only by lifeless stars and planets. Imagine the change in our worldview if we *knew* that living cells were metabolizing, reproducing, and dying on at least one exoplanet. If some of those cells belonged to intelligent life forms, the philosophical impact would be all the greater.

Summary of Common Misunderstandings

Given the current state of play in the search for life beyond the Earth, where as yet we have no conclusive evidence, it might seem inappropriate to discuss 'common misunderstandings'. But it's not. There can be misunderstandings about the way we search for extraterrestrial life, as well as in relation to the scientific basis for our search, and it is these that I focus on here, rather than misunderstandings about extraterrestrial life itself. I discuss them below in the order in which they're first encountered in the book.

Life can't be workably defined. On the contrary, a working definition that's useful as a basis in our search for extraterrestrial life is possible. It isn't perfect, but then few definitions are. The one I use here is called the RIM definition. This stands for Reproduction, Inheritance, and Metabolism. The full definition is given in Chapter 1.

The extent of the observable universe expands with improving technology. What expands with improving technology is the *observed* universe. This is much more extensive now than it was in Galileo's day. But the *observable* universe is the same today as it was then. Its limits are set by the universe's age and its (variable) rate of expansion. I give an estimate of its extent in Chapter 2.

The problem of a sample size of one is unsurmountable. Although we only have a sample of one inhabited planet – Earth – this doesn't mean that consideration of possible kinds of life elsewhere is a non-starter. On the contrary, various features of life on Earth help us to envisage constraints in terms of what's possible in general. These include the differences between organic and inorganic chemistry (pointing to carbon-based life elsewhere), and the near-universality of cells (pointing to cellular construction elsewhere). Also, the multiple occurrences of some evolutionary transitions on Earth – such as that from unicellular to multicellular – suggest that these are in some sense 'easy' and might be expected to occur also in the evolutionary processes of other planets. This argument, however, is complex, as I show in Chapter 3.

There's a definite galactic habitable zone.   The concept of a circumstellar habitable zone works well, at least for surface-based life, since it's defined by the possibility of having bodies of water on the surface of a planet (or moon) that are broadly equivalent to Earth's oceans and lakes. However, the idea of a *galactic* habitable zone doesn't work, because there is no precise criterion that distinguishes parts of a galaxy that are and aren't habitable. There may be probability variation for the existence of life across the galaxy, as I suggest in Chapter 4, but this is very different from a binary switch such as the existence or otherwise of liquid water.

There's definitely no life in the solar system apart from that on Earth.   Although there *may be* no life on other solar-system bodies, our exploration is not yet sufficiently comprehensive that we can say that there *is* no such life. The most unexplored of promising places in which life might exist are the subsurface oceans of several Jovian and Saturnian moons. It's unlikely that there are large multicellular life forms in these places – Europan whales, for example – but the possibility of microbial life in these bodies of water can't yet be ruled out, as I explain in Chapter 5.

Biogenic oxygen can't be distinguished from its photochemical counterpart.   Although some scientists describe oxygen as a robust biosignature gas, others emphasize the fact that it can be produced in various abiotic ways, including geochemical ones in or on rocks and photochemical ones in the atmosphere. While this is true, we shouldn't overstate the problem. If the first few planets discovered to have strong oxygen signatures are all found to be in habitable zones, then the likelihood is that the oxygen is biogenic, as I discuss on Chapter 6.

Life as we know it is a poor pointer to life in general.   In addition to the likelihood of parallel basics of life, such as carbon and cells, there may be parallels at a higher level, including parallels of our animal and plant kingdoms, with photosynthesis broadly characterizing one of these kingdoms, active movement the other. Parallel natural selection acting on organisms characterized by these features is likely to produce many similarities to Earthly organisms, including broad flat leaves in plants and bilateral symmetry in animals, as I explain in Chapter 7.

The Fermi paradox embodies a definite contradiction. A real or apparent contradiction is a key feature of what we call a paradox. Yet there may be no contradiction between the supposedly high probability of intelligent life on the one hand and the lack of any contact with it so far on the other, as embodied in the so-called Fermi paradox. Inhabited planets are likely to be common, but those inhabited by intelligent beings are probably a very small proportion of the total. Thus, while the nearest microbial life may be only a few light years away from Earth, the nearest intelligent life may be more than 100 times further, and the beings concerned may be focusing their search for life on planets orbiting their nearby stars, just as we do here, as I suggest in Chapter 8.

References and Further Reading

Chapter 1   The Search for Extraterrestrial Life

On the assessment of claims of the discovery of alien life

Green, J., Hoehler, T., Neveu, M., Domagal-Goldman, S., Scalice, D., and Voytek, M. 2021. Call for a framework for reporting evidence for life beyond Earth. *Nature*, 598: 575–579. https://doi.org/10.1038/s41586-021-03804-9.

On the life and astronomical work of Galileo Galilei

Livio, M. 2020. *Galileo and the Science Deniers*. Simon and Schuster, New York.

On telescopes

Cottrell, G. 2016. *Telescopes: A Very Short Introduction*. Oxford University Press, Oxford.

On the history of the search

Cocconi, G. and Morrison, P. 1959. Searching for interstellar communications. *Nature*, 184: 844–846.

Lowell, P. 1906. *Mars and Its Canals*. Macmillan, New York.

Lowell, P. 1908. *Mars as the Abode of Life*. Macmillan, New York.

Strous, L. 2020. Who discovered that the Sun was a star? Stanford Solar Center, Stanford, CA. http://solar-center.stanford.edu/FAQ/Qsunasstar.html.

On astrobiology in general

Catling, D. 2013. *Astrobiology: A Very Short Introduction*. Oxford University Press, Oxford.

Cockell, C. 2020. *Astrobiology: Understanding Life in the Universe*, 2nd edition. Wiley-Blackwell, Hoboken, NJ.

Kolb, V. (ed.) 2018. *Handbook of Astrobiology*. CRC Press, Boca Raton, FL.

Lingam, M. and Loeb, A. 2021. *Life in the Cosmos: From Biosignatures to Technosignatures*. Harvard University Press, Cambridge, MA.

Rothery, D. A., Gilmour, I., and Sephton, M. A. 2018. *An Introduction to Astrobiology*, 3rd edition. Cambridge University Press, Cambridge.

On intelligence

Cross, F. R., Carvell, G. E., Jackson, R. R., and Grace, R. C. 2020. Arthropod intelligence? The case for *Portia*. *Frontiers in Psychology*, 11. https://doi.org/10.3389/fpsyg.2020.568049.

Godfrey-Smith, P. 2017. *Other Minds: The Octopus and the Evolution of Intelligent Life*. Collins, London.

Neubauer, S., Hublin, J.-J., and Gunz, P. 2018. The evolution of modern human brain shape. *Science Advances*, 4: eaao5961. https://doi.org/10.1126/sciadv.aao5961.

On the idea that complex alien life is rare

Ward, P. D. and Brownlee, D. 2000. *Rare Earth: Why Complex Life Is Uncommon in the Universe*. Copernicus Books, New York.

Chapter 2   Where in the Universe to Look?

On understanding the nature of the universe

Natarajan, P. 2016. *Mapping the Heavens: The Radical Scientific Ideas that Reveal the Cosmos*. Yale University Press, New Haven, CT.

On galaxies

Jones, M. H., Lambourne, R. J. A., and Serjeant, S. (eds.) 2016. *An Introduction to Galaxies and Cosmology*, 2nd edition. Cambridge University Press, Cambridge.

On the discovery of the most distant galaxy to date

Harikane, Y. *et al.* (14 authors) 2022. A search for H-dropout Lyman break galaxies at z ~ 12–16. *The Astrophysical Journal*, 929: 1 (15 pp). https://doi.org/10.3847/1538-4357/ac53a9.

Oesch, P. A. *et al.* (18 authors) 2016. A remarkably luminous galaxy at z = 11.1 measured with Hubble Space Telescope grism technology. *The Astrophysical Journal*, 819: 129 (11 pp). https://doi.org/10.3847/0004-637X/819/2/129.

On the nature of stars

Green, S. F. and Jones, M. H. (eds.) 2015. *An Introduction to the Sun and Stars*, 2nd edition. Cambridge University Press, Cambridge.

On the possibility of life inside stars

Anchordoqui, L. and Chudnovsky, E. 2020. Can self-replicating species flourish in the interior of a star? *Letters in High Energy Physics*, LHEP-166. https://doi.org/10.31526/lhep.2020.166.

On the possibility of life on rogue planets

Stevenson, D. J. 1999. Life-sustaining planets in interstellar space? *Nature*, 400: 32.

Trefil, J. and Summers, M. 2019. *Imagined Life: A Speculative Scientific Journey among the Exoplanets in Search of Intelligent Aliens, Ice Creatures, and Supergravity Animals*. Smithsonian Books, Washington, DC.

Discovery of the Kepler 90 planetary system

Cabrera, J. *et al.* (10 authors) 2014. The planetary system to KIC 11442793: a compact analogue to the solar system. *The Astrophysical Journal*, 781: 18 (13 pp). https://doi.org/10.1088/0004-637X/781/1/18.

Shallue, C. J. and Vanderberg, A. 2018. Identifying planets with deep learning: a five-planet resonant chain around Kepler-80 and an eighth planet around Kepler-90. *The Astrophysical Journal*, 155: 94 (21 pp). https://doi.org/10.3847/1538-3881/aa9e09.

On the diversity of planets

Taylor, S. R. 2012. *Destiny or Chance Revisited: Planets and Their Place in the Cosmos*. Cambridge University Press, Cambridge.

On the formation of the first stars

Hashimoto, T. *et al.* (24 authors) 2018. The onset of star formation 250 million years after the Big Bang. *Nature*, 557: 392–395. https://doi.org/10.1038/s41586-018-0117-z.

Chapter 3   Evolution of Life – Here and Elsewhere

On the nature and origin of life

Pross, A. 2012. *What Is Life? How Chemistry Becomes Biology*. Oxford University Press, Oxford.

On the RNA world hypothesis

Bernhardt, H. S. 2012. The RNA world hypothesis: the worst theory of the early evolution of life (except for all the others)? *Biology Direct*, 7: 23 (10 pp). https://doi.org/10.1186/1745-6150-7-23.

Müller, F. *et al.* (9 authors) 2022. A prebiotically plausible scenario of an RNA-peptide world. *Nature*, 605: 279–284. https://doi.org/10.1038/s41586-022-04676-3.

Neveu, M., Kim, H.-J., and Benner, S. A. 2013. The 'strong' RNA world hypothesis: fifty years old. *Astrobiology*, 13: 391–403. https://doi.org/10.1089/ast.2012.0868.

On the properties of water

Finney, J. 2015. *Water: A Very Short Introduction*. Oxford University Press, Oxford.

On the classification of life on Earth

Doolittle, W. F. 2020. Evolution: two domains of life or three? *Current Biology*, 30: R177–R179. https://doi.org/10.1016/j.cub.2020.01.010.

Hennig, W. 1966. *Phylogenetic Systematics*. University of Illinois Press, Urbana, IL.

Whittaker, R. H. 1969. New concepts of kingdoms of organisms: evolutionary radiations are better represented by new classifications than by the traditional two kingdoms. *Science*, 163: 150–160.

Woese, C. R., Kandler, O., and Wheelis, M. L. 1990. Towards a natural system of organisms: proposal for the domains Archaea, Bacteria, and Eucarya. *Proceedings of the National Academy of Sciences of the USA*, 87: 4576–4579.

On the evolutionary origin of eukaryote cells

Martin, W. F., Garg, S., and Zimorski, V. 2015. Endosymbiotic theories for eukaryotic origin. *Philosophical Transactions of the Royal Society B*, 370: 20140330. https://doi.org/10.1098/rstb.2014.0330.

Satoh, N. 2020. *Endosymbiotic Theories of the Origins of Organelles Revisited: Retrospects and Prospects*. Springer Nature, Singapore.

Stadnichuk, I. N. and Kusnetsov, V. V. 2021. Endosymbiotic origin of chloroplasts in plant cells' evolution. *Russian Journal of Plant Physiology*, 68: 1–16. https://doi.org/10.1134/S1021443721010179.

On the role of animal movement in the evolution of intelligence

Wilkinson, M. 2016. *Restless Creatures: The Story of Life in Ten Movements*. Basic Books, New York.

Chapter 4 The Key Concept of Habitability
On the circumstellar habitable zone

Huang, S.-S. 1959. Occurrence of life in the Universe. *American Scientist*, 47: 397–402.

Kasting, J. 2010. *How to Find a Habitable Planet*. Princeton University Press, Princeton, NJ.

Kasting, J., Whitmire, D. P., and Reynolds, R. T. 1993. Habitable zones around main sequence stars. *Icarus*, 101: 108–128.

Ramirez, R. M. 2018. A more comprehensive habitable zone for finding life on other planets. *Geosciences*, 8: 280. https://doi.org/10.3390/geosciences8080280.

Tasker, E. 2017. *The Planet Factory: Exoplanets and the Search for a Second Earth*. Bloomsbury, London.

On the cause of the mass extinction event 66 million years ago

Schulte, P. *et al.* (42 authors) 2010. The Chicxulub asteroid impact and mass extinction at the Cretaceous–Paleogene boundary. *Science*, 327: 1214–1218. https://doi.org/10.1126/science.1177265.

On tidal locking of planets to their host stars

Barnes, R. 2017. Tidal locking of habitable exoplanets. *Celestial Mechanics and Dynamical Astronomy*, 129: 509–536. https://doi.org/10.1007/s10569-017-9783-7.

Pierrehumbert, R. T. and Hammond, M. 2019. Atmospheric circulation of tide-locked exoplanets. *Annual Review of Fluid Mechanics*, 51: 272–303. https://doi.org/10.1146/annurev-fluid-010518-040516.

Comprehensive information sources for exoplanets

Extrasolar Planets Encyclopaedia. http://exoplanet.eu.

NASA Exoplanet Archive. https://exoplanetarchive.ipac.caltech.edu.

Chapter 5 Life in the Solar System

On the assessment of claims of the discovery of alien life

Green, J., Hoehler, T., Neveu, M., Domagal-Goldman, S., Scalice, D., and Voytek, M. 2021. Call for a framework for reporting evidence for life beyond Earth. *Nature*, 598: 575–579. https://doi.org/10.1038/s41586-021-03804-9.

On the possibility of phosphine gas in the clouds of Venus

Greaves, J. S. *et al.* (19 authors) 2021. Phosphine gas in the cloud decks of Venus. *Nature Astronomy*, 5: 655–664. https://doi.org/10.1038/s41550-020-1174-4.

Villanueva, G. L. *et al.* (27 authors) 2021. No evidence of phosphine in the atmosphere of Venus from independent analyses. *Nature Astronomy*, 5: 631–635. https://doi.org/10.1038/s41550-021-01422-z.

On possible evidence for life on Mars

McKay, D. S. *et al.* (9 authors) 1996. Search for past life on Mars: possible relic biogenic activity in Martian meteorite ALH84001. *Science*, 273: 924–930. https://doi.org/10.1126/science.273.5277.924.

Misra, A. K., Acosta-Maeda, T. E., Scott, E. R. D., and Sharma, S. K. 2014. Possible mechanism for explaining the origin and size distribution of Martian hematite spherules. *Planetary and Space Science*, 92: 16–23. https://doi.org/10.1016/j.pss.2014.01.020.

On the moons of Jupiter and Saturn

Hansen, C. J. *et al.* (11 authors) 2011. The composition and structure of the Enceladus plume. *Geophysical Research Letters*, 38 (11). https://doi.org/10.1029/2011GL047415.

Hendrix, A. R. *et al.* (28 authors) 2019. The NASA roadmap to ocean worlds. *Astrobiology*, 19 (1): (27 pp). https://doi.org/10.1089/ast.2018.1955.

Mastrogiuseppe, M. *et al.* (7 authors) 2019. Deep and methane-rich lakes on Titan. *Nature Astronomy*, 3: 535–542. https://doi.org/10.1038/s41550-019-0714-2.

On the Breakthrough Initiatives

Merali, Z. 2015. News: Search for extra-terrestrial intelligence gets a $100 million boost – Russian billionaire Yuri Milner announces most comprehensive hunt for alien life. *Nature*, 523: 392–393. https://doi.org/10.1038/nature.2015.18016.

Milner, Y. *et al.* (32 authors) 2015. Open letter: Are we alone? Breakthrough Initiatives. https://breakthroughinitiatives.org/arewealone.

On the 2017 visitor to our solar system called 'Oumuamua

Loeb, A. 2021. *Extraterrestrial: The First Sign of Intelligent Life Beyond Earth*. John Murray, London.

Chapter 6   Life in Other Planetary Systems

Discoveries of exoplanets

Jenkins, J. M. *et al.* (29 authors) 2015. Discovery and validation of Kepler 452b: A 1.6 R_+ super Earth exoplanet in the habitable zone of a G2 star. *The Astronomical Journal*, 150: 56. https://doi.org/10.1088/0004-6256/150/2/56.

Mayor, M. and Queloz, D. 1995. A Jupiter-mass companion to a solar-type star. *Nature*, 378: 355–359.

Quintana, E. V. *et al.* (23 authors) 2014. An Earth-sized planet in the habitable zone of a cool star. *Science*, 344: 277–280. https://doi.org/10.1126/science.1249403.

Wolszczan, A. and Frail, D. A. 1992. A planetary system around the millisecond pulsar PSR1257 + 12. *Nature*, 355: 145–147.

On the TRAPPIST-1 planetary system

De Wit, J. *et al.* (17 authors) 2018. Atmospheric reconnaissance of the habitable-zone Earth-sized planets orbiting TRAPPIST-1. *Nature Astronomy*, 2: 214–219. https://doi.org/10.1038/s41550-017-0374-z.

Gillon, M. *et al.* (15 authors) 2016. Temperate Earth-sized planets transiting a nearby ultracool dwarf star. *Nature*, 533: 221–224. https://doi.org/10.1038/nature17448.

Gillon, M. *et al.* (30 authors) 2017. Seven temperate terrestrial planets around the nearby ultracool dwarf star TRAPPIST-1. *Nature*, 542: 456–460. https://doi.org/10.1038/nature21360.

On a variety of potentially habitable exoplanets

Anglada-Escudé, G. *et al.* (31 authors) 2016. A terrestrial planet candidate in a temperate orbit around Proxima Centauri. *Nature*, 536: 437–440. https://doi.org/10.1038/nature19106.

Feng, F. *et al.* (7 authors) 2017. Color difference makes a difference: four candidate planets around Tau Ceti. *The Astrophysical Journal*, 154: 135 (23 pp). https://doi .org/10.3847/1538-3881/aa83b4.

Gilbert, E. A. *et al.* (46 authors) 2020. The first habitable-zone Earth-sized planet from TESS: validation of the TOI-700 system. *The Astrophysical Journal*, 160: 116 (21 pp). https://doi.org/10.3847/1538-3881/aba4b3.

Jenkins, J. M. *et al.* (29 authors) 2015. Discovery and validation of Kepler 452b: A 1.6 R_+ super Earth exoplanet in the habitable zone of a G2 star. *The Astronomical Journal*, 150: 56 (19 pp). https://doi.org/10.1088/0004-6256/15 0/2/56.

Torres, G. *et al.* (19 authors) 2017. Validation of 12 small Kepler transiting planets in the habitable zone. *The Astrophysical Journal*, 154: 264 (19 pp). https://doi .org/10.3847/1538-3881/aa984b.

On exoplanet atmospheres

Catling, D. C. and Kasting, J. F. 2017. *Atmospheric Evolution on Inhabited and Lifeless Worlds*. Cambridge University Press, Cambridge.

Charbonneau, D., Brown, T. M., Noyes, R. W., and Gilliland, R. L. 2002. Detection of an extrasolar planet atmosphere. *The Astrophysical Journal*, 568: 377–384. https://doi.org/10.1086/338770.

Konopacky, Q. M., Barman, T. S., MacIntosh, B. A., and Marois, C. 2013. Detection of carbon monoxide and water absorption lines in an exoplanet atmosphere. *Science*, 339: 1398–1401. https://doi.org/10.1126/science .1232003.

Richardson, L. J., Deming, D., Horning, K., Seager, S., and Harrington, J. 2007. A spectrum of an extrasolar planet. *Nature*, 445: 892–895. https://doi.org/10 .1038/nature05636.

Schaefer, L. and Parmentier, V. 2021. The air over there: exploring exoplanet atmospheres. *Elements*, 17: 257–263. https://doi.org/10.2138/gselements .17.4.257.

Seager, S. 2010. *Exoplanet Atmospheres: Physical Processes*. Princeton University Press, Princeton and Oxford.

Chapter 7   The Nature of Extraterrestrial Life

On possible similarities between Earth life and alien life

Arthur, W. 2020. *The Biological Universe: Life in the Milky Way and Beyond*. Cambridge University Press, Cambridge.

Cockell, C. 2018. *The Equations of Life: How Physics Shapes Evolution*. Atlantic Books, London.

Conway Morris, S. 2003. *Life's Solution: Inevitable Humans in a Lonely Universe*. Cambridge University Press, Cambridge.

Darling, D. 2001. *Life Everywhere: The Maverick Science of Astrobiology*. Basic Books, New York.

Kershenbaum, A. 2020. *The Zoologist's Guide to the Galaxy: What Animals on Earth Reveal about Aliens – and Ourselves*. Penguin Random House, New York.

McGhee, G. R. 2019. *Convergent Evolution on Earth: Lessons for the Search for Extraterrestrial Life*. MIT Press, Cambridge, MA.

Schulze-Makuch, D. and Bains, W. 2017. *The Cosmic Zoo: Complex Life on Many Worlds*. Springer, Cham, Switzerland.

On possible differences between Earth life and alien life

Trefil, J. and Summers, M. 2019. *Imagined Life: A Speculative Scientific Journey Among the Exoplanets in Search of Intelligent Aliens, Ice Creatures, and Supergravity Animals*. Smithsonian Books, Washington DC.

Ward, P. D. and Brownlee, D. 2000. *Rare Earth: Why Complex Life Is Uncommon in the Universe*. Copernicus Books, New York.

On the survival of tardigrades in space

Jönnson, K. I., Rabbow, E., Schill, R. O., Harms-Ringdahl, R., and Rettberg, P. 2008. Tardigrades survive exposure to space in low Earth orbit. *Current Biology*, 18: R729–R731. https://doi.org/10.1016/j.cub.2008.06.048.

On natural selection

Darwin, C. 1859. *On the Origin of Species by Mean of Natural Selection, or the Preservation of Favoured Races in the Struggle for Life.* Murray, London.

Williams, G. C. 1992. *Natural Selection: Domains, Levels, and Challenges.* Oxford University Press, New York.

On the roles of systematic processes, history and chance in evolution

Conway Morris, S. 1998. *The Crucible of Creation: The Burgess Shale and the Rise of Animals.* Oxford University Press, Oxford.

Gould, S. J. 1983. *Hen's Teeth and Horse's Toes: Further Reflections in Natural History.* Norton, New York.

Gould, S. J. 1989. *Wonderful Life: The Burgess Shale and the Nature of History.* Norton, New York.

Monod, J. 1972. *Chance and Necessity: Essay on the Natural Philosophy of Modern Biology.* Knopf, New York.

Chapter 8 Intelligence – Here and Elsewhere

On human evolution

Almecija, S., Hammond, A. S., Thompson, N. E., Pugh, K. D., Moyà-Solà, S., and Alba, D. M. 2021. Fossil apes and human evolution. *Science*, 372 (6542). https://doi.org/10.1126/science.abb4363.

Harmand, S. *et al.* (21 authors) 2015. 3.3-million-year-old stone tools from Lomekwi-3, West Turkana, Kenya. *Nature*, 521: 310–315. https://doi.org/10.1038/nature14464.

Neubauer, S., Hublin, J.-J., and Gunz, P. 2018. The evolution of modern human brain shape. *Science Advances*, 4 (1): (8 pp). https://doi.org/10.1126/sciadv.aao5961.

Roberts, A. 2018. *Evolution: The Human Story*, 2nd edition. Dorling Kindersley, London.

Weaver, T. D. and Stringer, C. B. 2015. Unconstrained cranial evolution in Neandertals and modern humans compared to common chimpanzees. *Proceedings of the Royal Society B*, 282: 20151519. https://doi.org/10.1098/rs pb.2015.1519.

A fictional account of mobile predatory plants

Wyndham, J. 1951. *The Day of the Triffids*. Michael Joseph, London.

On the search for intelligent life via radio signals

Cocconi, G. and Morrison, P. 1959. Searching for interstellar communications. *Nature*, 184: 844–846.

Price, D. C. *et al.* (24 authors) 2020. The Breakthrough Listen search for intelligent life: observations of 1327 nearby stars over 1.10 to 3.45 GHz. *The Astronomical Journal*, 159: 86 (16 pp). https://doi.org/10.3847/1538-3881/ab65f1.

Sheikh, S. Z. *et al.* (18 authors) 2021. Analysis of the Breakthrough Listen signal of interest blc1 with a technosignature verification framework. *Nature Astronomy*, 5: 1153–1162. https://doi.org/10.1038/s41550-021-01508-8.

Traas, R. *et al.* (13 authors) 2021. The Breakthrough Listen search for intelligent life: searching for technosignatures in observations of TESS targets of interest. *The Astronomical Journal*, 161: 286 (12 pp). https://doi.org/10.3847/1538-3881/abf649.

On the Fermi paradox and the Drake equation

Drake, F. and Sobel, D. 1992. *Is Anyone Out There? The Scientific Search for Extraterrestrial Intelligence*. Bantam Doubleday Bell, New York.

Drake, F. and Sobel, D. 2010. The origin of the Drake equation. *Astronomy Beat*, 46: 1–4.

Webb, S. 2015. *If the Universe Is Teeming with Aliens ... Where Is Everybody? Seventy-Five Solutions to the Fermi Paradox and the Problem of Extraterrestrial Life*, 2nd edition. Springer International, Switzerland.

Figure Credits

A picture is often, as they say, worth a thousand words; that is, it tells a story that would otherwise take a considerable amount of print. The pictures in this book tell very different stories. Their spatial scales range from about a millimetre, for those tiny and almost indestructible animals called tardigrades, to about 100,000 light years, for the span of a galaxy such as the Milky Way. All but one of the illustrations herein are originals. They were commissioned specifically for the book, and the finished artwork was produced by Bridget Bravo. The exception is Figure 2.3, which has been modified from an illustration in *The Biological Universe* (Cambridge University Press, 2020).

Index

Page numbers in *italic* refer to figures

Abbreviations: SETI = search for extraterrestrial intelligence